Timed Addition Facts
Drills Improve Speed and Accuracy

Grades 1-3

Written by Ruth Solski
Illustrated by S&S Learning Materials

ISBN 978-1-55035-895-7
Copyright 2008

Published in the United States by:
On The Mark Press
3909 Witmer Road PMB 175
Niagara Falls, New York
14305
www.onthemarkpress.com

Published in Canada by:
S&S Learning Materials
15 Dairy Avenue
Napanee, Ontario
K7R 1M4
www.sslearning.com

At  Glance

Learning Expectations	Pages 4 to 6	Pages 7 to 10	Pages 11 to 14	Pages 15 to 18	Pages 19 to 22	Pages 23 to 24	Pages 25 to 28	Pages 29 to 32	Pages 33 to 36	Pages 37 to 40	Pages 41 to 42	Pages 43 to 46
Addition Facts												
• To strengthen addition fact recall	•	•	•	•	•	•	•	•	•	•	•	•
• Improve speed and accuracy in addition facts	•	•	•	•	•	•	•	•	•	•	•	•
• Develop the ability to memorize	•	•	•	•	•	•	•	•	•	•	•	•
Timed Drills												
• Zero Plus, One Plus	•											
• Two Plus		•										
• Three Plus			•									
• Four Plus				•								
• Five Plus					•							
• Zero Plus to Five Plus Reviews						•						
• Six Plus							•					
• Seven Plus								•				
• Eight Plus									•			
• Nine Plus										•		
• Six Plus to Nine Plus Review											•	
• Review Drills 0 to 18												•

Timed Addition Facts
Drills Improve Speed and Accuracy

Table of Contents

A Note to the Teacher:

The addition fact drills have been designed to help strengthen students' speed and accuracy through practice during a specified time or each student could be timed individually.

Some of the drills are shorter and are to be completed on a specified day. Each drill page concentrates on a specific area in addition fact recall. The drills proceed from the easiest level to the most difficult level. Each level has a daily practice page, a home practice page, and extra practice page, and a review test page.

The daily practice page is divided into five days. Each day of the week the student is to complete a drill. The date, score, and time it took are to be recorded in each section. This page could be glued into the student's workbooks or kept in individual file folders.

The home practice page is to be sent home to practice fact recall with parent supervision. Once completed it is to be returned to school signed by a parent. A letter of explanation should be sent home with the first home practice page explaining how it is to be completed.

The extra practice drill sheet is to be used with students who are still having difficulty recalling facts quickly and accurately. It is a different approach to the timed drill method. The student must supply the missing addend. There is no extra practice page for adding zero and adding one addition facts.

The review page or test page is to be used to test speed and accuracy within a given length of time. Begin with five minutes graduating down to one minute.

Tell students when to begin and when to stop. Have the student circle the last completed question with a red crayon or red pencil crayon. The students are to exchange their papers and to mark each incorrect answer with a red dot as you read the answers aloud. Have the student count the number of correct answers. No credit is to be given to incomplete answers. Record the number of correct answers, the time, and the date on each sheet where indicated. On each review test have all incomplete answers finished for extra practice.

There are two timed review tests for each section that may be used after each section has been practiced successfully. These tests will evaluate students' speed and accuracy in each section.

The final drill pages test all the facts from 0 to 18. These pages are to be used in the same manner as the other drills.

The results of the various drills may be recorded on the Score Record Sheets provided in the book.

Zero and One Plus Drills

Name: _____

Date: Monday_____   Score: _____/25   Time: _____ Min. _____ Sec.

0 + 0 = ____	6 + 1 = ____	8 + 1 = ____	0 + 5 = ____	5 + 1 = ____
1 + 1 = ____	0 + 3 = ____	0 + 3 = ____	1 + 6 = ____	2 + 0 = ____
3 + 1 = ____	2 + 1 = ____	9 + 0 = ____	4 + 1 = ____	1 + 4 = ____
0 + 5 = ____	2 + 0 = ____	1 + 7 = ____	0 + 1 = ____	0 + 7 = ____
2 + 1 = ____	9 + 1 = ____	2 + 1 = ____	8 + 0 = ____	8 + 1 = ____

Date: Tuesday_____   Score: _____/25   Time: _____ Min. _____ Sec.

9 + 1 = ____	3 + 1 = ____	5 + 0 = ____	0 + 8 = ____	2 + 0 = ____
8 + 1 = ____	0 + 5 = ____	0 + 9 = ____	8 + 1 = ____	1 + 5 = ____
6 + 0 = ____	9 + 0 = ____	1 + 6 = ____	1 + 0 = ____	3 + 1 = ____
0 + 4 = ____	8 + 1 = ____	9 + 1 = ____	7 + 1 = ____	0 + 9 = ____
1 + 2 = ____	5 + 1 = ____	0 + 6 = ____	5 + 0 = ____	8 + 0 = ____

Date: Wednesday_____   Score: _____/25   Time: _____ Min. _____ Sec.

8 + 1 = ____	1 + 1 = ____	2 + 1 = ____	9 + 0 = ____	8 + 0 = ____
6 + 1 = ____	1 + 3 = ____	0 + 7 = ____	1 + 7 = ____	7 + 1 = ____
4 + 1 = ____	1 + 5 = ____	1 + 3 = ____	8 + 1 = ____	0 + 0 = ____
2 + 1 = ____	1 + 7 = ____	6 + 1 = ____	0 + 3 = ____	6 + 1 = ____
1 + 9 = ____	1 + 0 = ____	9 + 1 = ____	5 + 1 = ____	0 + 8 = ____

Date: Thursday_____   Score: _____/25   Time: _____ Min. _____ Sec.

0 + 9 = ____	3 + 1 = ____	6 + 1 = ____	0 + 9 = ____	1 + 9 = ____
8 + 1 = ____	1 + 2 = ____	7 + 0 = ____	9 + 1 = ____	1 + 0 = ____
5 + 1 = ____	0 + 4 = ____	0 + 3 = ____	0 + 5 = ____	5 + 1 = ____
1 + 7 = ____	1 + 6 = ____	2 + 1 = ____	4 + 1 = ____	0 + 3 = ____
0 + 0 = ____	1 + 0 = ____	1 + 7 = ____	8 + 1 = ____	4 + 1 = ____

Date: Friday_____   Score: _____/25   Time: _____ Min. _____ Sec.

8 + 1 = ____	6 + 0 = ____	9 + 1 = ____	1 + 7 = ____	9 + 0 = ____
1 + 3 = ____	1 + 5 = ____	1 + 4 = ____	5 + 1 = ____	8 + 1 = ____
0 + 7 = ____	7 + 1 = ____	1 + 8 = ____	0 + 6 = ____	7 + 0 = ____
1 + 1 = ____	0 + 8 = ____	5 + 0 = ____	1 + 2 = ____	1 + 4 = ____
0 + 4 = ____	9 + 0 = ____	1 + 0 = ____	6 + 1 = ____	2 + 1 = ____

Home Practice Zero and One Plus Drills

Name: _____

Monday	Tuesday	Wednesday	Thursday	Friday
0 + 0 = ___	1 + 0 = ___	9 + 1 = ___	8 + 1 = ___	7 + 1 = ___
3 + 1 = ___	1 + 2 = ___	9 + 0 = ___	1 + 9 = ___	8 + 0 = ___
0 + 4 = ___	1 + 8 = ___	0 + 5 = ___	8 + 0 = ___	9 + 1 = ___
1 + 6 = ___	9 + 1 = ___	2 + 1 = ___	1 + 6 = ___	7 + 0 = ___
7 + 0 = ___	4 + 0 = ___	1 + 5 = ___	7 + 0 = ___	1 + 6 = ___
1 + 9 = ___	1 + 7 = ___	0 + 9 = ___	4 + 1 = ___	1 + 1 = ___
6 + 1 = ___	2 + 0 = ___	1 + 6 = ___	2 + 0 = ___	0 + 2 = ___
3 + 0 = ___	0 + 1 = ___	4 + 0 = ___	1 + 7 = ___	3 + 1 = ___
1 + 5 = ___	2 + 0 = ___	3 + 1 = ___	1 + 5 = ___	0 + 6 = ___
5 + 0 = ___	1 + 2 = ___	6 + 0 = ___	0 + 7 = ___	5 + 1 = ___
1 + 8 = ___	4 + 1 = ___	1 + 8 = ___	1 + 8 = ___	9 + 0 = ___
5 + 1 = ___	0 + 5 = ___	0 + 9 = ___	1 + 0 = ___	9 + 1 = ___
0 + 0 = ___	6 + 0 = ___	7 + 1 = ___	0 + 8 = ___	0 + 7 = ___
1 + 1 = ___	1 + 7 = ___	1 + 9 = ___	6 + 1 = ___	2 + 1 = ___
9 + 1 = ___	9 + 1 = ___	0 + 1 = ___	0 + 2 = ___	0 + 3 = ___
1 + 5 = ___	1 + 6 = ___	1 + 1 = ___	1 + 2 = ___	3 + 1 = ___
9 + 0 = ___	4 + 0 = ___	6 + 1 = ___	7 + 1 = ___	1 + 2 = ___
1 + 2 = ___	0 + 7 = ___	0 + 9 = ___	0 + 8 = ___	0 + 7 = ___
4 + 1 = ___	2 + 1 = ___	1 + 5 = ___	5 + 1 = ___	8 + 1 = ___
0 + 3 = ___	3 + 0 = ___	0 + 2 = ___	3 + 0 = ___	1 + 5 = ___
1 + 2 = ___	0 + 5 = ___	9 + 1 = ___	8 + 1 = ___	7 + 1 = ___
7 + 1 = ___	1 + 5 = ___	1 + 5 = ___	6 + 0 = ___	1 + 4 = ___
8 + 0 = ___	9 + 1 = ___	7 + 0 = ___	0 + 9 = ___	0 + 5 = ___
2 + 1 = ___	0 + 0 = ___	1 + 9 = ___	3 + 1 = ___	1 + 7 = ___
6 + 1 = ___	6 + 0 = ___	0 + 4 = ___	0 + 8 = ___	0 + 9 = ___
Score: ____/25 _____ Min. _____ Sec.	Score: ____/25 _____ Min. _____ Sec.	Score: ____/25 _____ Min. _____ Sec.	Score: ____/25 _____ Min. _____ Sec.	Score: ____/25 _____ Min. _____ Sec.

Zero and One Plus Review Test

Name: _____

1 + 2	1 + 0	0 + 0	9 + 1	1 + 8	9 + 0	1 + 5	3 + 1	5 + 1	0 + 7
1 + 4	2 + 1	7 + 1	0 + 0	1 + 3	1 + 8	1 + 1	0 + 2	4 + 1	8 + 1
4 + 0	0 + 2	1 + 1	5 + 1	0 + 4	2 + 1	1 + 2	9 + 1	1 + 7	8 + 0
9 + 0	4 + 1	2 + 1	1 + 1	1 + 8	0 + 0	1 + 4	7 + 1	2 + 0	1 + 2
1 + 8	9 + 1	0 + 1	3 + 1	1 + 5	0 + 7	3 + 0	1 + 3	8 + 1	0 + 3
1 + 4	0 + 3	1 + 9	7 + 1	5 + 1	0 + 1	1 + 9	1 + 4	6 + 1	5 + 0
8 + 1	6 + 1	1 + 5	1 + 2	0 + 4	9 + 1	1 + 1	6 + 1	0 + 2	1 + 7
0 + 6	1 + 8	7 + 1	0 + 1	5 + 1	1 + 3	6 + 0	1 + 2	1 + 9	3 + 1
1 + 2	0 + 3	1 + 0	6 + 1	1 + 7	4 + 1	1 + 9	7 + 0	0 + 6	1 + 5
1 + 6	5 + 1	0 + 9	8 + 1	1 + 3	4 + 0	0 + 4	9 + 1	1 + 7	4 + 1

Date: _____   Score: _____ /100   Time: _____ Min. _____ Sec.

Two Plus Drills

Name: _____

Date: Monday_____   Score: _____/25 Time: _____ Min. _____ Sec.

2 + 3 = ___	2 + 2 = ___	2 + 6 = ___	2 + 1 = ___	2 + 4 = ___
2 + 5 = ___	2 + 4 = ___	2 + 4 = ___	2 + 6 = ___	2 + 0 = ___
2 + 8 = ___	2 + 7 = ___	2 + 0 = ___	2 + 9 = ___	2 + 8 = ___
2 + 2 = ___	2 + 3 = ___	2 + 5 = ___	2 + 2 = ___	2 + 7 = ___
2 + 4 = ___	2 + 5 = ___	2 + 2 = ___	2 + 7 = ___	2 + 9 = ___

Date: Tuesday_____   Score: _____/25 Time: _____ Min. _____ Sec.

2 + 0 = ___	2 + 1 = ___	2 + 7 = ___	2 + 5 = ___	2 + 2 = ___
2 + 6 = ___	2 + 8 = ___	2 + 9 = ___	2 + 3 = ___	2 + 6 = ___
2 + 9 = ___	2 + 0 = ___	2 + 3 = ___	2 + 8 = ___	2 + 4 = ___
2 + 7 = ___	2 + 9 = ___	2 + 8 = ___	2 + 4 = ___	2 + 1 = ___
2 + 1 = ___	2 + 6 = ___	2 + 1 = ___	2 + 0 = ___	2 + 5 = ___

Date: Wednesday _____   Score: _____/25 Time: _____ Min. _____ Sec.

2 + 3 = ___	2 + 2 = ___	2 + 6 = ___	2 + 1 = ___	2 + 0 = ___
2 + 5 = ___	2 + 4 = ___	2 + 4 = ___	2 + 6 = ___	2 + 3 = ___
2 + 8 = ___	2 + 7 = ___	2 + 0 = ___	2 + 9 = ___	2 + 8 = ___
2 + 2 = ___	2 + 3 = ___	2 + 2 = ___	2 + 2 = ___	2 + 7 = ___
2 + 4 = ___	2 + 5 = ___	2 + 8 = ___	2 + 7 = ___	2 + 9 = ___

Date: Thursday _____   Score: _____/25 Time: _____ Min. _____ Sec.

2 + 0 = ___	2 + 1 = ___	2 + 0 = ___	2 + 5 = ___	2 + 2 = ___
2 + 6 = ___	2 + 8 = ___	2 + 5 = ___	2 + 3 = ___	2 + 6 = ___
2 + 9 = ___	2 + 0 = ___	2 + 2 = ___	2 + 8 = ___	2 + 4 = ___
2 + 1 = ___	2 + 6 = ___	2 + 7 = ___	2 + 4 = ___	2 + 1 = ___
2 + 3 = ___	2 + 2 = ___	2 + 9 = ___	2 + 0 = ___	2 + 5 = ___

Date: Friday _____   Score: _____/25 Time: _____ Min. _____ Sec.

2 + 5 = ___	2 + 4 = ___	2 + 3 = ___	2 + 1 = ___	2 + 0 = ___
2 + 8 = ___	2 + 7 = ___	2 + 8 = ___	2 + 6 = ___	2 + 3 = ___
2 + 2 = ___	2 + 3 = ___	2 + 1 = ___	2 + 9 = ___	2 + 8 = ___
2 + 4 = ___	2 + 5 = ___	2 + 6 = ___	2 + 2 = ___	2 + 7 = ___
2 + 9 = ___	2 + 1 = ___	2 + 4 = ___	2 + 7 = ___	2 + 9 = ___

Home Practice Two Plus Drills

Name: _____

Monday	Tuesday	Wednesday	Thursday	Friday
2 + 2 = ____	2 + 4 = ____	2 + 6 = ____	2 + 1 = ____	2 + 3 = ____
2 + 4 = ____	2 + 0 = ____	2 + 4 = ____	2 + 6 = ____	2 + 5 = ____
2 + 7 = ____	2 + 8 = ____	2 + 0 = ____	2 + 9 = ____	2 + 8 = ____
2 + 3 = ____	2 + 7 = ____	2 + 5 = ____	2 + 2 = ____	2 + 2 = ____
2 + 5 = ____	2 + 9 = ____	2 + 2 = ____	2 + 7 = ____	2 + 4 = ____
2 + 1 = ____	2 + 2 = ____	2 + 7 = ____	2 + 5 = ____	2 + 0 = ____
2 + 8 = ____	2 + 6 = ____	2 + 9 = ____	2 + 3 = ____	2 + 6 = ____
2 + 0 = ____	2 + 4 = ____	2 + 3 = ____	2 + 8 = ____	2 + 9 = ____
2 + 9 = ____	2 + 1 = ____	2 + 8 = ____	2 + 4 = ____	2 + 7 = ____
2 + 6 = ____	2 + 5 = ____	2 + 1 = ____	2 + 0 = ____	2 + 1 = ____
2 + 2 = ____	2 + 0 = ____	2 + 6 = ____	2 + 1 = ____	2 + 3 = ____
2 + 4 = ____	2 + 3 = ____	2 + 4 = ____	2 + 6 = ____	2 + 5 = ____
2 + 7 = ____	2 + 8 = ____	2 + 0 = ____	2 + 9 = ____	2 + 8 = ____
2 + 3 = ____	2 + 7 = ____	2 + 2 = ____	2 + 2 = ____	2 + 2 = ____
2 + 5 = ____	2 + 9 = ____	2 + 4 = ____	2 + 7 = ____	2 + 4 = ____
2 + 1 = ____	2 + 2 = ____	2 + 0 = ____	2 + 5 = ____	2 + 0 = ____
2 + 8 = ____	2 + 6 = ____	2 + 5 = ____	2 + 3 = ____	2 + 6 = ____
2 + 0 = ____	2 + 4 = ____	2 + 2 = ____	2 + 8 = ____	2 + 9 = ____
2 + 6 = ____	2 + 1 = ____	2 + 7 = ____	2 + 4 = ____	2 + 1 = ____
2 + 2 = ____	2 + 5 = ____	2 + 9 = ____	2 + 0 = ____	2 + 3 = ____
2 + 4 = ____	2 + 0 = ____	2 + 3 = ____	2 + 1 = ____	2 + 5 = ____
2 + 7 = ____	2 + 3 = ____	2 + 8 = ____	2 + 6 = ____	2 + 8 = ____
2 + 3 = ____	2 + 8 = ____	2 + 1 = ____	2 + 9 = ____	2 + 2 = ____
2 + 5 = ____	2 + 7 = ____	2 + 6 = ____	2 + 2 = ____	2 + 4 = ____
2 + 1 = ____	2 + 9 = ____	2 + 4 = ____	2 + 7 = ____	2 + 9 = ____
Score: ____/25	Score: ____/25	Score: ____/25	Score: ____/25	Score: ____/25
_____ Min.	_____ Min.	_____ Min.	_____ Min.	_____ Min.
_____ Sec.	_____ Sec.	_____ Sec.	_____ Sec.	_____ Sec.

Extra Practice Two Plus Drills

Name: _____

Day 1	Day 2	Day 3	Day 4	Day 5
2 + ___ = 4	2 + ___ = 6	___ + 6 = 8	2 + ___ = 5	2 + ___ = 5
___ + 4 = 6	2 + ___ = 2	2 + ___ = 6	___ + 6 = 8	___ + 5 = 7
2 + ___ = 9	___ + 8 =10	___ + 0 = 2	2 + ___ =11	2 + ___ =10
___ + 3 = 5	2 + ___ = 9	2 + ___ = 7	___ + 2 = 4	___ + 2 = 4
___ + 5 = 7	2 + ___ =11	___ + 2 = 4	2 + ___ = 9	2 + ___ = 6
2 + ___ = 3	___ + 2 = 4	2 + ___ = 9	___ + 5 = 7	___ + 0 = 2
2 + ___ =10	2 + ___ = 8	___ + 9 =11	2 + ___ = 5	2 + ___ = 8
___ + 0 = 2	___ + 4 = 6	2 + ___ = 5	___ + 8 =10	2 + ___ =11
2 + ___ =11	___ + 1 = 3	___ + 8 =10	2 + ___ = 6	___ + 7 = 9
___ + 6 = 8	2 + ___ = 7	2 + ___ = 3	___ + 0 = 2	2 + ___ = 3
2 + ___ = 4	2 + ___ = 2	___ + 6 = 8	2 + ___ = 3	___ + 3 = 5
2 + ___ = 6	2 + ___ = 5	2 + ___ = 6	___ + 6 = 8	2 + ___ = 7
___ + 7 = 9	2 + ___ =10	___ + 0 = 2	2 + 9 = ___	___ + 8 =10
2 + ___ = 5	2 + ___ = 9	2 + ___ = 4	___ + 2 = 4	2 + ___ = 4
___ + 5 = 7	___ + 9 =11	___ + 4 = 6	2 + ___ = 9	___ + 4 = 6
2 + ___ = 3	___ + 2 = 4	2 + ___ = 2	___ + 5 = 7	2 + ___ = 2
2 + ___ =10	2 + ___ = 8	___ + 5 = 7	2 + ___ = 5	___ + 6 = 8
2 + ___ = 2	2 + ___ = 6	2 + ___ = 4	___ + 8 =10	2 + ___ =11
___ + 6 = 8	___ + 1 = 3	___ + 7 = 9	2 + ___ = 6	___ + 1 = 3
2 + ___ = 4	2 + ___ = 7	2 + ___ =11	2 + ___ = 2	2 + ___ = 5
2 + ___ = 6	___ + 0 = 2	___ + 3 = 5	___ + 1 = 3	___ + 5 = 7
___ + 7 = 9	2 + ___ = 5	2 + ___ =10	2 + ___ = 8	2 + ___ =10
2 + ___ = 5	2 + ___ = 8	___ + 1 = 3	___ + 9 =11	___ + 2 = 4
___ + 5 = 7	2 + ___ = 9	2 + ___ = 8	2 + ___ = 4	2 + ___ = 6
2 + ___ = 3	2 + ___ =11	___ + 4 = 6	___ + 7 = 9	___ + 9 =11
Score: _____/25	Score: _____/25	Score: _____/25	Score: _____/25	Score: _____/25
_____ Min.	_____ Min.	_____ Min.	_____ Min.	_____ Min.
_____ Sec.	_____ Sec.	_____ Sec.	_____ Sec.	_____ Sec.

Two Plus Review Test

Name: _____

6 + 2	3 + 2	1 + 2	4 + 2	8 + 2	5 + 2	0 + 2	9 + 2	2 + 2	7 + 2
4 + 2	2 + 2	7 + 2	0 + 2	8 + 2	1 + 2	5 + 2	3 + 2	2 + 2	6 + 2
3 + 2	1 + 2	7 + 2	9 + 2	4 + 2	2 + 2	5 + 2	8 + 2	6 + 2	0 + 2
3 + 2	2 + 2	0 + 2	5 + 2	9 + 2	1 + 2	7 + 2	6 + 2	4 + 2	8 + 2
6 + 2	9 + 2	3 + 2	2 + 2	8 + 2	1 + 2	4 + 2	0 + 2	5 + 2	7 + 2
1 + 2	6 + 2	8 + 2	5 + 2	4 + 2	3 + 2	7 + 2	9 + 2	2 + 2	0 + 2
5 + 2	3 + 2	2 + 2	1 + 2	4 + 2	0 + 2	9 + 2	3 + 2	6 + 2	8 + 2
9 + 2	5 + 2	2 + 2	7 + 2	4 + 2	8 + 2	3 + 2	6 + 2	1 + 2	2 + 2
3 + 2	6 + 2	0 + 2	4 + 2	8 + 2	1 + 2	7 + 2	5 + 2	9 + 2	2 + 2
7 + 2	9 + 2	4 + 2	0 + 2	6 + 2	8 + 2	2 + 2	3 + 2	1 + 2	5 + 2

Date: _____ **Score:** _____/100 **Time:** _____ Min. _____ Sec.

Three Plus Drills

Name: _____

Date: Monday_____ Score: _____ /25 Time: _____ Min. _____ Sec.

3 + 3 = ____	3 + 2 = ____	3 + 6 = ____	3 + 1 = ____	3 + 4 = ____
3 + 5 = ____	3 + 4 = ____	3 + 4 = ____	3 + 6 = ____	3 + 0 = ____
3 + 8 = ____	3 + 7 = ____	3 + 0 = ____	3 + 9 = ____	3 + 8 = ____
3 + 2 = ____	3 + 3 = ____	3 + 5 = ____	3 + 2 = ____	3 + 7 = ____
3 + 4 = ____	3 + 5 = ____	3 + 2 = ____	3 + 7 = ____	3 + 9 = ____

Date: Tuesday_____ Score: _____ /25 Time: _____ Min. _____ Sec.

3 + 0 = ____	3 + 1 = ____	3 + 7 = ____	3 + 5 = ____	3 + 2 = ____
3 + 6 = ____	3 + 8 = ____	3 + 9 = ____	3 + 3 = ____	3 + 6 = ____
3 + 9 = ____	3 + 0 = ____	3 + 3 = ____	3 + 8 = ____	3 + 4 = ____
3 + 7 = ____	3 + 9 = ____	3 + 8 = ____	3 + 4 = ____	3 + 1 = ____
3 + 1 = ____	3 + 6 = ____	3 + 1 = ____	3 + 0 = ____	3 + 5 = ____

Date: Wednesday _____ Score: _____ /25 Time: _____ Min. _____ Sec.

3 + 3 = ____	3 + 2 = ____	3 + 6 = ____	3 + 1 = ____	3 + 0 = ____
3 + 5 = ____	3 + 4 = ____	3 + 4 = ____	3 + 6 = ____	3 + 3 = ____
3 + 8 = ____	3 + 7 = ____	3 + 0 = ____	3 + 9 = ____	3 + 8 = ____
3 + 2 = ____	3 + 3 = ____	3 + 2 = ____	3 + 2 = ____	3 + 5 = ____
3 + 4 = ____	3 + 5 = ____	3 + 4 = ____	3 + 7 = ____	3 + 9 = ____

Date: Thursday _____ Score: _____ /25 Time: _____ Min. _____ Sec.

3 + 0 = ____	3 + 1 = ____	3 + 0 = ____	3 + 5 = ____	3 + 2 = ____
3 + 6 = ____	3 + 8 = ____	3 + 5 = ____	3 + 3 = ____	3 + 6 = ____
3 + 9 = ____	3 + 0 = ____	3 + 2 = ____	3 + 8 = ____	3 + 4 = ____
3 + 1 = ____	3 + 6 = ____	3 + 7 = ____	3 + 4 = ____	3 + 1 = ____
3 + 3 = ____	3 + 2 = ____	3 + 9 = ____	3 + 0 = ____	3 + 5 = ____

Date: Friday_____ Score: _____ /25 Time: _____ Min. _____ Sec.

3 + 5 = ____	3 + 4 = ____	3 + 3 = ____	3 + 1 = ____	3 + 0 = ____
3 + 8 = ____	3 + 7 = ____	3 + 8 = ____	3 + 6 = ____	3 + 3 = ____
3 + 2 = ____	3 + 3 = ____	3 + 1 = ____	3 + 9 = ____	3 + 8 = ____
3 + 4 = ____	3 + 5 = ____	3 + 6 = ____	3 + 2 = ____	3 + 7 = ____
3 + 9 = ____	3 + 1 = ____	3 + 4 = ____	3 + 7 = ____	3 + 9 = ____

Home Practice Three Plus Drills

Name: _____

Monday	Tuesday	Wednesday	Thursday	Friday
3 + 6 = ____	3 + 2 = ____	3 + 4 = ____	3 + 3 = ____	3 + 1 = ____
3 + 4 = ____	3 + 4 = ____	3 + 0 = ____	3 + 5 = ____	3 + 6 = ____
3 + 0 = ____	3 + 7 = ____	3 + 8 = ____	3 + 8 = ____	3 + 9 = ____
3 + 5 = ____	3 + 3 = ____	3 + 7 = ____	3 + 2 = ____	3 + 2 = ____
3 + 2 = ____	3 + 5 = ____	3 + 9 = ____	3 + 4 = ____	3 + 7 = ____
3 + 7 = ____	3 + 1 = ____	3 + 2 = ____	3 + 0 = ____	3 + 5 = ____
3 + 9 = ____	3 + 8 = ____	3 + 6 = ____	3 + 6 = ____	3 + 3 = ____
3 + 3 = ____	3 + 0 = ____	3 + 4 = ____	3 + 9 = ____	3 + 8 = ____
3 + 8 = ____	3 + 9 = ____	3 + 1 = ____	3 + 7 = ____	3 + 4 = ____
3 + 1 = ____	3 + 6 = ____	3 + 5 = ____	3 + 1 = ____	3 + 0 = ____
3 + 6 = ____	3 + 2 = ____	3 + 0 = ____	3 + 3 = ____	3 + 1 = ____
3 + 4 = ____	3 + 4 = ____	3 + 3 = ____	3 + 5 = ____	3 + 6 = ____
3 + 0 = ____	3 + 7 = ____	3 + 8 = ____	3 + 8 = ____	3 + 9 = ____
3 + 2 = ____	3 + 3 = ____	3 + 5 = ____	3 + 2 = ____	3 + 2 = ____
3 + 4 = ____	3 + 5 = ____	3 + 9 = ____	3 + 4 = ____	3 + 7 = ____
3 + 0 = ____	3 + 1 = ____	3 + 2 = ____	3 + 0 = ____	3 + 5 = ____
3 + 5 = ____	3 + 8 = ____	3 + 6 = ____	3 + 6 = ____	3 + 3 = ____
3 + 2 = ____	3 + 0 = ____	3 + 4 = ____	3 + 9 = ____	3 + 8 = ____
3 + 7 = ____	3 + 6 = ____	3 + 1 = ____	3 + 1 = ____	3 + 4 = ____
3 + 9 = ____	3 + 2 = ____	3 + 5 = ____	3 + 3 = ____	3 + 0 = ____
3 + 3 = ____	3 + 4 = ____	3 + 0 = ____	3 + 5 = ____	3 + 1 = ____
3 + 8 = ____	3 + 7 = ____	3 + 3 = ____	3 + 8 = ____	3 + 6 = ____
3 + 1 = ____	3 + 3 = ____	3 + 8 = ____	3 + 2 = ____	3 + 9 = ____
3 + 6 = ____	3 + 5 = ____	3 + 7 = ____	3 + 4 = ____	3 + 2 = ____
3 + 4 = ____	3 + 1 = ____	3 + 9 = ____	3 + 9 = ____	3 + 7 = ____
Score: ____ /25 ____ Min. ____ Sec.	Score: ____ /25 ____ Min. ____ Sec.	Score: ____ /25 ____ Min. ____ Sec.	Score: ____ /25 ____ Min. ____ Sec.	Score: ____ /25 ____ Min. ____ Sec.

Extra Practice Three Plus Drills

Name: _____

Day 1	Day 2	Day 3	Day 4	Day 5
3 + ___ = 6	3 + ___ = 5	___ + 6 = 9	3 + ___ = 4	___ + 4 = 7
___ + 5 = 8	___ + 4 = 7	3 + ___ = 7	___ + 6 = 9	3 + ___ = 3
3 + ___ = 11	3 + ___ = 10	___ + 0 = 3	3 + ___ = 12	___ + 8 = 11
3 + ___ = 5	___ + 3 = 6	3 + ___ = 8	___ + 2 = 5	3 + ___ = 10
___ + 4 = 7	3 + ___ = 8	___ + 2 = 5	3 + ___ = 10	___ + 9 = 12
3 + ___ = 3	___ + 1 = 4	3 + ___ = 10	___ + 5 = 8	3 + ___ = 5
___ + 6 = 9	3 + ___ = 11	___ + 9 = 12	3 + ___ = 6	___ + 6 = 9
3 + ___ = 3	___ + 0 = 3	3 + ___ = 6	___ + 8 = 11	3 + ___ = 7
___ + 7 = 10	3 + ___ = 12	___ + 8 = 11	3 + ___ = 7	___ + 1 = 4
3 + ___ = 4	___ + 6 = 9	3 + ___ = 4	___ + 0 = 3	3 + ___ = 8
___ + 3 = 6	3 + ___ = 5	___ + 6 = 9	3 + ___ = 4	___ + 0 = 3
3 + ___ = 8	___ + 4 = 7	3 + ___ = 7	___ + 6 = 9	3 + ___ = 6
___ + 8 = 11	3 + ___ = 10	___ + 0 = 3	3 + ___ = 9	___ + 8 = 11
3 + ___ = 5	___ + 3 = 6	3 + ___ = 5	___ + 2 = 5	3 + ___ = 8
___ + 4 = 7	3 + ___ = 8	___ + 4 = 7	3 + ___ = 10	___ + 9 = 12
3 + ___ = 3	___ + 1 = 4	3 + ___ = 3	___ + 5 = 8	3 + ___ = 5
___ + 6 = 9	3 + ___ = 11	___ + 5 = 8	3 + ___ = 6	___ + 6 = 9
3 + ___ = 12	___ + 0 = 3	3 + ___ = 5	___ + 8 = 11	3 + ___ = 7
___ + 1 = 4	3 + ___ = 9	___ + 7 = 10	3 + ___ = 7	___ + 1 = 4
3 + ___ = 6	___ + 2 = 5	3 + ___ = 12	___ + 0 = 3	3 + ___ = 8
___ + 5 = 8	3 + ___ = 7	___ + 3 = 6	3 + ___ = 4	___ + 9 = 12
3 + ___ = 11	___ + 7 = 10	3 + ___ = 11	___ + 6 = 9	3 + ___ = 5
___ + 2 = 6	3 + ___ = 6	___ + 1 = 3	3 + ___ = 12	___ + 6 = 9
3 + ___ = 7	___ + 5 = 8	3 + ___ = 9	___ + 2 = 5	3 + ___ = 7
___ + 9 = 12	3 + ___ = 4	___ + 4 = 7	3 + ___ = 10	___ + 1 = 4
Score: _____/25	Score: _____/25	Score: _____/25	Score: _____/25	Score: _____/25
_____ Min.	_____ Min.	_____ Min.	_____ Min.	_____ Min.
_____ Sec.	_____ Sec.	_____ Sec.	_____ Sec.	_____ Sec.

Three Plus Review Test

Name: _____

3 + 3	5 + 3	8 + 3	2 + 3	4 + 3	0 + 3	6 + 3	9 + 3	7 + 3	1 + 3
5 + 3	1 + 3	9 + 3	6 + 3	0 + 3	4 + 3	2 + 3	8 + 3	5 + 3	3 + 3
8 + 3	3 + 3	2 + 3	4 + 3	9 + 3	2 + 3	4 + 3	7 + 3	3 + 3	5 + 3
1 + 3	8 + 3	0 + 3	9 + 3	6 + 3	2 + 3	4 + 3	7 + 3	1 + 3	3 + 3
5 + 3	1 + 3	8 + 3	0 + 3	6 + 3	2 + 3	7 + 3	4 + 3	7 + 3	5 + 3
1 + 3	5 + 3	6 + 3	4 + 3	3 + 3	0 + 3	5 + 3	2 + 3	7 + 3	9 + 3
3 + 3	8 + 3	1 + 3	6 + 3	4 + 3	2 + 3	0 + 3	4 + 3	0 + 3	5 + 3
2 + 3	7 + 3	9 + 3	3 + 3	8 + 3	1 + 3	6 + 3	1 + 3	6 + 3	4 + 3
9 + 3	2 + 3	7 + 3	5 + 3	3 + 3	8 + 3	4 + 3	0 + 3	1 + 3	6 + 3
2 + 3	0 + 3	9 + 3	4 + 3	6 + 3	7 + 3	1 + 3	5 + 3	8 + 3	3 + 3

Date: _____   Score: _____/100   Time: _____ Min. _____ Sec.

OTM-1139 • SSK1-39 Timed Addition Facts

Four Plus Drills

Name: _____

Date: Monday _____ Score: ____/25 Time: ____ Min. ____ Sec.

4 + 3 = ___	4 + 2 = ___	4 + 6 = ___	4 + 1 = ___	4 + 4 = ___
4 + 5 = ___	4 + 4 = ___	4 + 4 = ___	4 + 6 = ___	4 + 0 = ___
4 + 8 = ___	4 + 7 = ___	4 + 0 = ___	4 + 9 = ___	4 + 8 = ___
4 + 2 = ___	4 + 3 = ___	4 + 5 = ___	4 + 2 = ___	4 + 7 = ___
4 + 4 = ___	4 + 5 = ___	4 + 2 = ___	4 + 7 = ___	4 + 9 = ___

Date: Tuesday _____ Score: ____/25 Time: ____ Min. ____ Sec.

4 + 0 = ___	4 + 1 = ___	4 + 7 = ___	4 + 5 = ___	4 + 2 = ___
4 + 6 = ___	4 + 8 = ___	4 + 9 = ___	4 + 3 = ___	4 + 6 = ___
4 + 9 = ___	4 + 0 = ___	4 + 3 = ___	4 + 8 = ___	4 + 4 = ___
4 + 7 = ___	4 + 9 = ___	4 + 8 = ___	4 + 4 = ___	4 + 1 = ___
4 + 1 = ___	4 + 6 = ___	4 + 1 = ___	4 + 0 = ___	4 + 5 = ___

Date: Wednesday _____ Score: ____/25 Time: ____ Min. ____ Sec.

4 + 3 = ___	4 + 2 = ___	4 + 6 = ___	4 + 1 = ___	4 + 0 = ___
4 + 5 = ___	4 + 4 = ___	4 + 4 = ___	4 + 6 = ___	4 + 3 = ___
4 + 8 = ___	4 + 7 = ___	4 + 0 = ___	4 + 9 = ___	4 + 8 = ___
4 + 2 = ___	4 + 3 = ___	4 + 2 = ___	4 + 2 = ___	4 + 5 = ___
4 + 4 = ___	4 + 5 = ___	4 + 4 = ___	4 + 7 = ___	4 + 9 = ___

Date: Thursday _____ Score: ____/25 Time: ____ Min. ____ Sec.

4 + 0 = ___	4 + 1 = ___	4 + 0 = ___	4 + 5 = ___	4 + 2 = ___
4 + 6 = ___	4 + 8 = ___	4 + 5 = ___	4 + 3 = ___	4 + 6 = ___
4 + 9 = ___	4 + 0 = ___	4 + 2 = ___	4 + 8 = ___	4 + 4 = ___
4 + 1 = ___	4 + 6 = ___	4 + 7 = ___	4 + 4 = ___	4 + 1 = ___
4 + 3 = ___	4 + 2 = ___	4 + 9 = ___	4 + 0 = ___	4 + 5 = ___

Date: Friday _____ Score: ____/25 Time: ____ Min. ____ Sec.

4 + 5 = ___	4 + 4 = ___	4 + 3 = ___	4 + 1 = ___	4 + 0 = ___
4 + 8 = ___	4 + 7 = ___	4 + 8 = ___	4 + 6 = ___	4 + 3 = ___
4 + 2 = ___	4 + 3 = ___	4 + 1 = ___	4 + 9 = ___	4 + 8 = ___
4 + 4 = ___	4 + 5 = ___	4 + 6 = ___	4 + 2 = ___	4 + 7 = ___
4 + 9 = ___	4 + 1 = ___	4 + 4 = ___	4 + 7 = ___	4 + 9 = ___

Home Practice Four Plus Drills

Name: _____

Monday	Tuesday	Wednesday	Thursday	Friday
4 + 3 = ___	4 + 2 = ___	4 + 6 = ___	4 + 1 = ___	4 + 4 = ___
4 + 5 = ___	4 + 4 = ___	4 + 4 = ___	4 + 6 = ___	4 + 0 = ___
4 + 8 = ___	4 + 7 = ___	4 + 0 = ___	4 + 9 = ___	4 + 8 = ___
4 + 2 = ___	4 + 3 = ___	4 + 5 = ___	4 + 2 = ___	4 + 7 = ___
4 + 4 = ___	4 + 5 = ___	4 + 2 = ___	4 + 7 = ___	4 + 9 = ___
4 + 0 = ___	4 + 1 = ___	4 + 7 = ___	4 + 5 = ___	4 + 2 = ___
4 + 6 = ___	4 + 8 = ___	4 + 9 = ___	4 + 3 = ___	4 + 6 = ___
4 + 9 = ___	4 + 0 = ___	4 + 3 = ___	4 + 8 = ___	4 + 4 = ___
4 + 7 = ___	4 + 9 = ___	4 + 8 = ___	4 + 4 = ___	4 + 1 = ___
4 + 1 = ___	4 + 6 = ___	4 + 1 = ___	4 + 0 = ___	4 + 5 = ___
4 + 3 = ___	4 + 2 = ___	4 + 6 = ___	4 + 1 = ___	4 + 0 = ___
4 + 5 = ___	4 + 4 = ___	4 + 4 = ___	4 + 6 = ___	4 + 3 = ___
4 + 8 = ___	4 + 7 = ___	4 + 0 = ___	4 + 9 = ___	4 + 8 = ___
4 + 2 = ___	4 + 3 = ___	4 + 2 = ___	4 + 2 = ___	4 + 5 = ___
4 + 4 = ___	4 + 5 = ___	4 + 4 = ___	4 + 7 = ___	4 + 9 = ___
4 + 0 = ___	4 + 1 = ___	4 + 0 = ___	4 + 5 = ___	4 + 2 = ___
4 + 6 = ___	4 + 8 = ___	4 + 5 = ___	4 + 3 = ___	4 + 6 = ___
4 + 9 = ___	4 + 0 = ___	4 + 2 = ___	4 + 8 = ___	4 + 4 = ___
4 + 1 = ___	4 + 6 = ___	4 + 7 = ___	4 + 4 = ___	4 + 1 = ___
4 + 3 = ___	4 + 2 = ___	4 + 9 = ___	4 + 0 = ___	4 + 5 = ___
4 + 5 = ___	4 + 4 = ___	4 + 3 = ___	4 + 1 = ___	4 + 0 = ___
4 + 8 = ___	4 + 7 = ___	4 + 8 = ___	4 + 6 = ___	4 + 3 = ___
4 + 2 = ___	4 + 3 = ___	4 + 1 = ___	4 + 9 = ___	4 + 8 = ___
4 + 4 = ___	4 + 5 = ___	4 + 6 = ___	4 + 2 = ___	4 + 7 = ___
4 + 9 = ___	4 + 1 = ___	4 + 4 = ___	4 + 7 = ___	4 + 9 = ___
Score: ____/25 ____ Min. ____ Sec.	Score: ____/25 ____ Min. ____ Sec.	Score: ____/25 ____ Min. ____ Sec.	Score: ____/25 ____ Min. ____ Sec.	Score: ____/25 ____ Min. ____ Sec.

OTM-1139 • SSK1-39 Timed Addition Facts

Extra Practice Four Plus Drills

Name: _____

Day 1	Day 2	Day 3	Day 4	Day 5
4 + ___ = 7	___ + 2 = 6	4 + ___ = 10	___ + 4 = 5	4 + ___ = 8
___ + 5 = 9	4 + ___ = 8	___ + 4 = 8	4 + ___ = 10	___ + 0 = 4
4 + ___ = 12	___ + 7 = 11	4 + ___ = 4	___ + 9 = 13	4 + ___ = 12
___ + 2 = 6	4 + ___ = 7	___ + 5 = 9	4 + ___ = 6	___ + 7 = 11
4 + ___ = 8	___ + 5 = 9	4 + ___ = 6	___ + 7 = 11	4 + ___ = 13
___ + 0 = 4	4 + ___ = 5	___ + 7 = 11	4 + ___ = 9	___ + 2 = 6
4 + ___ = 10	___ + 8 = 12	4 + ___ = 13	___ + 3 = 7	4 + ___ = 10
___ + 9 = 13	4 + ___ = 10	___ + 3 = 7	4 + ___ = 12	___ + 4 = 8
4 + ___ = 11	___ + 9 = 13	4 + ___ = 12	___ + 4 = 8	4 + ___ = 5
___ + 1 = 5	4 + ___ = 10	___ + 1 = 5	4 + ___ = 4	___ + 5 = 9
4 + ___ = 7	___ + 2 = 6	4 + ___ = 10	___ + 1 = 5	4 + ___ = 4
___ + 5 = 9	4 + ___ = 8	___ + 4 = 8	4 + ___ = 10	___ + 3 = 7
4 + ___ = 12	___ + 7 = 11	4 + ___ = 4	___ + 9 = 13	4 + ___ = 12
___ + 2 = 6	4 + ___ = 7	___ + 2 = 6	4 + ___ = 6	___ + 5 = 9
4 + ___ = 8	___ + 5 = 9	4 + ___ = 8	___ + 7 = 11	4 + ___ = 13
___ + 0 = 4	4 + ___ = 5	___ + 0 = 4	4 + ___ = 9	___ + 2 = 6
4 + ___ = 10	___ + 8 = 12	4 + ___ = 9	___ + 3 = 7	4 + ___ = 10
___ + 9 = 13	4 + ___ = 4	___ + 2 = 6	4 + ___ = 12	___ + 4 = 8
4 + ___ = 5	___ + 6 = 10	4 + ___ = 11	___ + 4 = 8	4 + ___ = 5
___ + 3 = 7	4 + ___ = 6	___ + 9 = 13	4 + ___ = 4	___ + 5 = 9
4 + ___ = 9	___ + 4 = 8	4 + ___ = 7	___ + 1 = 5	4 + ___ = 4
___ + 8 = 12	4 + ___ = 11	___ + 8 = 12	4 + ___ = 10	___ + 3 = 7
4 + ___ = 6	___ + 3 = 7	4 + ___ = 5	___ + 9 = 12	4 + ___ = 12
___ + 4 = 8	4 + ___ = 9	___ + 6 = 10	4 + ___ = 6	___ + 7 = 11
4 + ___ = 13	___ + 1 = 5	4 + ___ = 8	___ + 7 = 11	4 + ___ = 13
Score: _____/25	Score: _____/25	Score: _____/25	Score: _____/25	Score: _____/25
_____ Min.	_____ Min.	_____ Min.	_____ Min.	_____ Min.
_____ Sec.	_____ Sec.	_____ Sec.	_____ Sec.	_____ Sec.

Four Plus Review Test

Name: _____

3 +4	5 +4	8 +4	2 +4	4 +4	0 +4	6 +4	9 +4	7 +4	1 +4
3 +4	5 +4	8 +4	2 +4	0 +4	4 +4	1 +4	3 +4	6 +4	9 +4
5 +4	8 +4	2 +4	4 +4	9 +4	2 +4	4 +4	7 +4	3 +4	5 +4
1 +4	0 +4	9 +4	6 +4	8 +4	7 +4	3 +4	5 +4	2 +4	4 +4
8 +4	2 +4	7 +4	1 +4	4 +4	3 +4	0 +4	1 +4	6 +4	5 +4
6 +4	0 +4	4 +4	2 +4	3 +4	1 +4	5 +4	7 +4	8 +4	9 +4
4 +4	2 +4	0 +4	6 +4	5 +4	0 +4	4 +4	2 +4	7 +4	9 +4
3 +4	8 +4	1 +4	9 +4	2 +4	6 +4	7 +4	4 +4	5 +4	1 +4
3 +4	0 +4	8 +4	6 +4	4 +4	2 +4	1 +4	9 +4	7 +4	4 +4
0 +4	8 +4	7 +4	9 +4	2 +4	6 +4	4 +4	1 +4	5 +4	0 +4

Date: _____ Score: _____/100 Time: _____ Min. _____ Sec.

Five Plus Drills

Name: _____

Date: Monday_____ Score: _____/25 Time: _____ Min. _____ Sec.

5 + 3 = ____	5 + 2 = ____	5 + 6 = ____	5 + 1 = ____	5 + 0 = ____
5 + 5 = ____	5 + 4 = ____	5 + 4 = ____	5 + 6 = ____	5 + 0 = ____
5 + 8 = ____	5 + 7 = ____	5 + 0 = ____	5 + 9 = ____	5 + 8 = ____
5 + 2 = ____	5 + 3 = ____	5 + 5 = ____	5 + 2 = ____	5 + 7 = ____
5 + 4 = ____	5 + 5 = ____	5 + 2 = ____	5 + 7 = ____	5 + 9 = ____

Date: Tuesday_____ Score: _____/25 Time: _____ Min. _____ Sec.

5 + 0 = ____	5 + 1 = ____	5 + 7 = ____	5 + 5 = ____	5 + 2 = ____
5 + 6 = ____	5 + 8 = ____	5 + 9 = ____	5 + 3 = ____	5 + 6 = ____
5 + 9 = ____	5 + 0 = ____	5 + 3 = ____	5 + 8 = ____	5 + 4 = ____
5 + 7 = ____	5 + 9 = ____	5 + 8 = ____	5 + 4 = ____	5 + 1 = ____
5 + 4 = ____	5 + 6 = ____	5 + 1 = ____	5 + 0 = ____	5 + 5 = ____

Date: Wednesday _____ Score: _____/25 Time: _____ Min. _____ Sec.

5 + 5 = ____	5 + 4 = ____	5 + 3 = ____	5 + 1 = ____	5 + 0 = ____
5 + 8 = ____	5 + 7 = ____	5 + 8 = ____	5 + 6 = ____	5 + 3 = ____
5 + 2 = ____	5 + 3 = ____	5 + 1 = ____	5 + 9 = ____	5 + 8 = ____
5 + 4 = ____	5 + 5 = ____	5 + 6 = ____	5 + 2 = ____	5 + 7 = ____
5 + 9 = ____	5 + 1 = ____	5 + 4 = ____	5 + 7 = ____	5 + 9 = ____

Date: Thursday _____ Score: _____/25 Time: _____ Min. _____ Sec.

5 + 0 = ____	5 + 1 = ____	5 + 0 = ____	5 + 5 = ____	5 + 2 = ____
5 + 6 = ____	5 + 8 = ____	5 + 5 = ____	5 + 3 = ____	5 + 6 = ____
5 + 9 = ____	5 + 0 = ____	5 + 2 = ____	5 + 8 = ____	5 + 4 = ____
5 + 1 = ____	5 + 6 = ____	5 + 7 = ____	5 + 4 = ____	5 + 1 = ____
5 + 3 = ____	5 + 2 = ____	5 + 9 = ____	5 + 0 = ____	5 + 5 = ____

Date: Friday_____ Score: _____/25 Time: _____ Min. _____ Sec.

5 + 3 = ____	5 + 2 = ____	5 + 6 = ____	5 + 1 = ____	5 + 0 = ____
5 + 5 = ____	5 + 4 = ____	5 + 4 = ____	5 + 6 = ____	5 + 3 = ____
5 + 8 = ____	5 + 7 = ____	5 + 0 = ____	5 + 9 = ____	5 + 8 = ____
5 + 2 = ____	5 + 3 = ____	5 + 2 = ____	5 + 2 = ____	5 + 5 = ____
5 + 4 = ____	5 + 5 = ____	5 + 4 = ____	5 + 7 = ____	5 + 9 = ____

Home Practice Five Plus Drills

Name: _____

Monday	Tuesday	Wednesday	Thursday	Friday
5 + 0 = ___	5 + 6 = ___	5 + 1 = ___	5 + 2 = ___	5 + 3 = ___
5 + 0 = ___	5 + 4 = ___	5 + 6 = ___	5 + 4 = ___	5 + 5 = ___
5 + 8 = ___	5 + 0 = ___	5 + 9 = ___	5 + 7 = ___	5 + 8 = ___
5 + 7 = ___	5 + 5 = ___	5 + 2 = ___	5 + 3 = ___	5 + 2 = ___
5 + 9 = ___	5 + 2 = ___	5 + 7 = ___	5 + 5 = ___	5 + 4 = ___
5 + 2 = ___	5 + 7 = ___	5 + 5 = ___	5 + 1 = ___	5 + 0 = ___
5 + 6 = ___	5 + 9 = ___	5 + 3 = ___	5 + 8 = ___	5 + 6 = ___
5 + 4 = ___	5 + 3 = ___	5 + 8 = ___	5 + 0 = ___	5 + 9 = ___
5 + 1 = ___	5 + 8 = ___	5 + 4 = ___	5 + 9 = ___	5 + 7 = ___
5 + 5 = ___	5 + 1 = ___	5 + 0 = ___	5 + 6 = ___	5 + 4 = ___
5 + 0 = ___	5 + 6 = ___	5 + 1 = ___	5 + 2 = ___	5 + 3 = ___
5 + 3 = ___	5 + 4 = ___	5 + 6 = ___	5 + 4 = ___	5 + 5 = ___
5 + 8 = ___	5 + 0 = ___	5 + 9 = ___	5 + 7 = ___	5 + 8 = ___
5 + 8 = ___	5 + 2 = ___	5 + 2 = ___	5 + 3 = ___	5 + 2 = ___
5 + 9 = ___	5 + 4 = ___	5 + 7 = ___	5 + 5 = ___	5 + 4 = ___
5 + 2 = ___	5 + 0 = ___	5 + 5 = ___	5 + 1 = ___	5 + 0 = ___
5 + 6 = ___	5 + 5 = ___	5 + 3 = ___	5 + 8 = ___	5 + 6 = ___
5 + 4 = ___	5 + 2 = ___	5 + 8 = ___	5 + 0 = ___	5 + 9 = ___
5 + 1 = ___	5 + 7 = ___	5 + 4 = ___	5 + 6 = ___	5 + 1 = ___
5 + 5 = ___	5 + 9 = ___	5 + 0 = ___	5 + 2 = ___	5 + 3 = ___
5 + 0 = ___	5 + 3 = ___	5 + 1 = ___	5 + 4 = ___	5 + 5 = ___
5 + 3 = ___	5 + 8 = ___	5 + 6 = ___	5 + 7 = ___	5 + 8 = ___
5 + 8 = ___	5 + 1 = ___	5 + 9 = ___	5 + 3 = ___	5 + 2 = ___
5 + 7 = ___	5 + 6 = ___	5 + 2 = ___	5 + 5 = ___	5 + 4 = ___
5 + 9 = ___	5 + 4 = ___	5 + 7 = ___	5 + 1 = ___	5 + 9 = ___
Score: _____/25 _____ Min. _____ Sec.	Score: _____/25 _____ Min. _____ Sec.	Score: _____/25 _____ Min. _____ Sec.	Score: _____/25 _____ Min. _____ Sec.	Score: _____/25 _____ Min. _____ Sec.

OTM-1139 • SSK1-39 Timed Addition Facts

Extra Practice Five Plus Drills

Name: _____

Day 1	Day 2	Day 3	Day 4	Day 5
5 + ___ = 8	___ + 2 = 7	5 + ___ = 11	___ + 1 = 6	5 + ___ = 5
___ + 5 = 10	5 + ___ = 9	___ + 4 = 9	5 + ___ = 11	___ + 8 = 13
5 + ___ = 13	___ + 7 = 12	5 + ___ = 5	___ + 9 = 14	___ + 0 = 5
___ + 2 = 7	5 + ___ = 8	___ + 5 = 10	5 + ___ = 7	5 + ___ = 12
5 + ___ = 9	___ + 5 = 10	5 + ___ = 7	___ + 7 = 12	___ + 9 = 13
___ + 0 = 5	5 + ___ = 6	___ + 7 = 12	5 + ___ = 10	5 + ___ = 7
5 + ___ = 11	___ + 8 = 13	5 + ___ = 13	___ + 3 = 8	___ + 6 = 11
___ + 9 = 14	5 + ___ = 5	___ + 3 = 8	5 + ___ = 13	5 + ___ = 9
5 + ___ = 12	___ + 9 = 14	5 + ___ = 13	___ + 4 = 9	___ + 1 = 6
___ + 1 = 6	5 + ___ = 11	___ + 1 = 6	5 + ___ = 5	5 + ___ = 10
5 + ___ = 8	___ + 2 = 7	5 + ___ = 11	___ + 1 = 6	___ + 0 = 5
___ + 5 = 10	5 + ___ = 9	___ + 4 = 10	5 + ___ = 11	5 + ___ = 8
5 + ___ = 13	___ + 7 = 12	5 + ___ = 5	___ + 9 = 14	___ + 8 = 13
___ + 2 = 7	5 + ___ = 8	___ + 2 = 7	5 + ___ = 7	5 + ___ = 10
5 + ___ = 9	___ + 5 = 10	5 + ___ = 9	___ + 7 = 9	___ + 9 = 14
___ + 0 = 5	5 + ___ = 6	___ + 0 = 5	5 + ___ = 10	5 + ___ = 7
5 + ___ = 11	___ + 8 = 13	5 + ___ = 10	___ + 3 = 8	___ + 6 = 11
___ + 9 = 14	5 + ___ = 5	___ + 2 = 7	5 + ___ = 13	5 + ___ = 9
5 + ___ = 6	___ + 6 = 11	5 + ___ = 9	___ + 4 = 9	___ + 1 = 6
___ + 3 = 8	5 + ___ = 7	___ + 9 = 13	5 + ___ = 5	5 + ___ = 10
5 + ___ = 10	___ + 4 = 9	5 + ___ = 8	___ + 1 = 6	___ + 0 = 5
___ + 8 = 13	5 + ___ = 12	___ + 8 = 13	5 + ___ = 11	5 + ___ = 8
5 + ___ = 7	___ + 3 = 8	5 + ___ = 6	___ + 9 = 14	___ + 8 = 13
___ + 4 = 9	5 + ___ = 10	___ + 6 = 11	5 + ___ = 7	5 + ___ = 12
5 + ___ = 14	___ + 1 = 6	5 + ___ = 9	5 + ___ = 12	___ + 9 = 14
Score: _____ /25	Score: _____ /25	Score: _____ /25	Score: _____ /25	Score: _____ /25
_____ Min.	_____ Min.	_____ Min.	_____ Min.	_____ Min.
_____ Sec.	_____ Sec.	_____ Sec.	_____ Sec.	_____ Sec.

Five Plus Review Test

Name: _____

3 +5	5 +5	8 +5	2 +5	4 +5	0 +5	6 +5	9 +5	7 +5	1 +5
3 +5	8 +5	2 +5	5 +5	6 +5	9 +5	4 +5	1 +5	0 +5	3 +5
8 +5	5 +5	2 +5	4 +5	5 +5	2 +5	4 +5	9 +5	7 +5	5 +5
3 +5	7 +5	4 +5	2 +5	6 +5	9 +5	0 +5	8 +5	3 +5	1 +5
5 +5	1 +5	8 +5	0 +5	2 +5	4 +5	6 +5	7 +5	3 +5	5 +5
1 +5	6 +5	4 +5	5 +5	7 +5	0 +5	2 +5	9 +5	8 +5	3 +5
1 +5	3 +5	9 +5	4 +5	5 +5	1 +5	8 +5	7 +5	2 +5	1 +5
6 +5	8 +5	2 +5	0 +5	7 +5	6 +5	1 +5	2 +5	4 +5	6 +5
9 +5	4 +5	7 +5	1 +5	3 +5	9 +5	6 +5	9 +5	0 +5	4 +5
2 +5	0 +5	5 +5	6 +5	8 +5	2 +5	4 +5	3 +5	5 +5	0 +5

Date: _____   Score: _____/100   Time: _____ Min. _____ Sec.

+0, 1+, 2+, 3+, 4+, 5+ Timed Drill Review

Name: _____

Row 1	Row 2	Row 3	Row 4
2 + 3 = _____	3 + 2 = _____	4 + 6 = _____	5 + 1 = _____
3 + 5 = _____	4 + 4 = _____	5 + 4 = _____	2 + 6 = _____
4 + 8 = _____	5 + 7 = _____	2 + 0 = _____	3 + 9 = _____
5 + 2 = _____	2 + 3 = _____	3 + 5 = _____	4 + 2 = _____
2 + 4 = _____	3 + 5 = _____	4 + 2 = _____	5 + 7 = _____
3 + 0 = _____	4 + 1 = _____	5 + 7 = _____	2 + 5 = _____
4 + 6 = _____	5 + 8 = _____	2 + 9 = _____	3 + 3 = _____
5 + 9 = _____	2 + 0 = _____	3 + 3 = _____	4 + 8 = _____
2 + 7 = _____	3 + 9 = _____	4 + 8 = _____	5 + 4 = _____
3 + 1 = _____	4 + 6 = _____	5 + 1 = _____	2 + 0 = _____
4 + 3 = _____	5 + 2 = _____	2 + 6 = _____	3 + 1 = _____
5 + 5 = _____	2 + 4 = _____	3 + 4 = _____	4 + 6 = _____
2 + 8 = _____	3 + 7 = _____	4 + 0 = _____	5 + 9 = _____
3 + 2 = _____	4 + 3 = _____	5 + 2 = _____	2 + 2 = _____
4 + 4 = _____	5 + 5 = _____	2 + 4 = _____	3 + 7 = _____
5 + 0 = _____	2 + 1 = _____	3 + 0 = _____	4 + 5 = _____
2 + 6 = _____	3 + 8 = _____	4 + 5 = _____	5 + 3 = _____
3 + 9 = _____	4 + 0 = _____	5 + 2 = _____	2 + 8 = _____
4 + 1 = _____	5 + 6 = _____	2 + 7 = _____	3 + 4 = _____
5 + 3 = _____	2 + 2 = _____	3 + 9 = _____	4 + 0 = _____
2 + 5 = _____	3 + 4 = _____	4 + 3 = _____	5 + 1 = _____
3 + 8 = _____	4 + 7 = _____	5 + 8 = _____	2 + 6 = _____
4 + 2 = _____	5 + 3 = _____	2 + 1 = _____	3 + 9 = _____
5 + 4 = _____	2 + 5 = _____	3 + 6 = _____	4 + 2 = _____
2 + 9 = _____	3 + 1 = _____	4 + 4 = _____	5 + 7 = _____

Date: _____   Score: _____/100   Time: _____ Min. _____ Sec.

+0, 1+, 2+, 3+, 4+, 5+ Timed Drill Review

Name: _____

3 +2	3 +5	4 +8	5 +2	2 +4	3 +0	4 +6	5 +9	2 +7	3 +1
4 +3	5 +5	2 +8	3 +2	4 +4	5 +0	2 +6	3 +9	4 +1	5 +3
2 +5	3 +8	4 +2	5 +4	2 +9	3 +2	4 +4	5 +7	2 +3	3 +5
4 +1	5 +8	2 +0	3 +9	4 +6	5 +2	2 +4	3 +7	4 +3	5 +5
2 +1	3 +8	4 +0	5 +6	2 +2	3 +4	4 +7	5 +3	2 +5	3 +1
4 +6	5 +4	2 +0	3 +5	4 +2	5 +7	2 +9	3 +3	4 +8	5 +1
2 +6	3 +4	4 +0	5 +2	2 +4	3 +0	4 +5	5 +2	2 +7	3 +9
4 +3	5 +8	2 +1	3 +6	4 +4	5 +1	2 +6	3 +9	4 +2	5 +7
2 +5	3 +3	4 +8	5 +4	2 +0	3 +1	4 +6	5 +9	2 +2	3 +7
4 +5	5 +3	2 +8	3 +4	4 +0	5 +1	2 +6	3 +9	4 +2	5 +7

Date: _____ **Score:** _____ /100 **Time:** _____ Min. _____ Sec.

OTM-1139 • SSK1-39 Timed Addition Facts

Six Plus Drills

Name: _____

Date: Monday_____ Score: _____ /25   Time: _____ Min. _____ Sec.

6 + 5 = ___	6 + 4 = ___	6 + 3 = ___	6 + 1 = ___	6 + 0 = ___
6 + 8 = ___	6 + 7 = ___	6 + 8 = ___	6 + 6 = ___	6 + 3 = ___
6 + 2 = ___	6 + 3 = ___	6 + 1 = ___	6 + 9 = ___	6 + 8 = ___
6 + 4 = ___	6 + 5 = ___	6 + 6 = ___	6 + 2 = ___	6 + 7 = ___
6 + 9 = ___	6 + 1 = ___	6 + 4 = ___	6 + 7 = ___	6 + 9 = ___

Date: Tuesday_____ Score: _____ /25   Time: _____ Min. _____ Sec.

6 + 0 = ___	6 + 1 = ___	6 + 0 = ___	6 + 5 = ___	6 + 2 = ___
6 + 6 = ___	6 + 8 = ___	6 + 5 = ___	6 + 3 = ___	6 + 6 = ___
6 + 9 = ___	6 + 0 = ___	6 + 2 = ___	6 + 8 = ___	6 + 4 = ___
6 + 4 = ___	6 + 6 = ___	6 + 7 = ___	6 + 4 = ___	6 + 1 = ___
6 + 3 = ___	6 + 2 = ___	6 + 9 = ___	6 + 0 = ___	6 + 5 = ___

Date: Wednesday _____ Score: _____ /25   Time: _____ Min. _____ Sec.

6 + 3 = ___	6 + 2 = ___	6 + 6 = ___	6 + 1 = ___	6 + 0 = ___
6 + 5 = ___	6 + 4 = ___	6 + 4 = ___	6 + 6 = ___	6 + 3 = ___
6 + 8 = ___	6 + 7 = ___	6 + 0 = ___	6 + 9 = ___	6 + 8 = ___
6 + 2 = ___	6 + 3 = ___	6 + 2 = ___	6 + 2 = ___	6 + 5 = ___
6 + 4 = ___	6 + 5 = ___	6 + 4 = ___	6 + 7 = ___	6 + 9 = ___

Date: Thursday _____ Score: _____ /25   Time: _____ Min. _____ Sec.

6 + 0 = ___	6 + 1 = ___	6 + 7 = ___	6 + 5 = ___	6 + 2 = ___
6 + 6 = ___	6 + 8 = ___	6 + 9 = ___	6 + 3 = ___	6 + 6 = ___
6 + 9 = ___	6 + 0 = ___	6 + 3 = ___	6 + 8 = ___	6 + 4 = ___
6 + 7 = ___	6 + 9 = ___	6 + 8 = ___	6 + 4 = ___	6 + 1 = ___
6 + 1 = ___	6 + 6 = ___	6 + 1 = ___	6 + 0 = ___	6 + 5 = ___

Date: Friday_____ Score: _____ /25   Time: _____ Min. _____ Sec.

6 + 3 = ___	6 + 2 = ___	6 + 6 = ___	6 + 1 = ___	6 + 4 = ___
6 + 5 = ___	6 + 4 = ___	6 + 4 = ___	6 + 6 = ___	6 + 0 = ___
6 + 8 = ___	6 + 7 = ___	6 + 0 = ___	6 + 9 = ___	6 + 8 = ___
6 + 2 = ___	6 + 3 = ___	6 + 5 = ___	6 + 2 = ___	6 + 7 = ___
6 + 4 = ___	6 + 5 = ___	6 + 2 = ___	6 + 7 = ___	6 + 9 = ___

Home Practice Six Plus Drills

Name: _____

Monday	Tuesday	Wednesday	Thursday	Friday
6 + 6 = ____	6 + 3 = ____	6 + 1 = ____	6 + 2 = ____	6 + 4 = ____
6 + 4 = ____	6 + 5 = ____	6 + 6 = ____	6 + 4 = ____	6 + 0 = ____
6 + 0 = ____	6 + 8 = ____	6 + 9 = ____	6 + 7 = ____	6 + 8 = ____
6 + 5 = ____	6 + 2 = ____	6 + 2 = ____	6 + 3 = ____	6 + 7 = ____
6 + 2 = ____	6 + 4 = ____	6 + 7 = ____	6 + 5 = ____	6 + 9 = ____
6 + 7 = ____	6 + 0 = ____	6 + 5 = ____	6 + 1 = ____	6 + 2 = ____
6 + 9 = ____	6 + 6 = ____	6 + 3 = ____	6 + 8 = ____	6 + 6 = ____
6 + 3 = ____	6 + 9 = ____	6 + 8 = ____	6 + 0 = ____	6 + 4 = ____
6 + 8 = ____	6 + 7 = ____	6 + 4 = ____	6 + 9 = ____	6 + 1 = ____
6 + 1 = ____	6 + 1 = ____	6 + 0 = ____	6 + 6 = ____	6 + 5 = ____
6 + 6 = ____	6 + 3 = ____	6 + 1 = ____	6 + 2 = ____	6 + 0 = ____
6 + 4 = ____	6 + 5 = ____	6 + 6 = ____	6 + 4 = ____	6 + 3 = ____
6 + 0 = ____	6 + 8 = ____	6 + 9 = ____	6 + 7 = ____	6 + 8 = ____
6 + 2 = ____	6 + 2 = ____	6 + 2 = ____	6 + 3 = ____	6 + 5 = ____
6 + 4 = ____	6 + 4 = ____	6 + 7 = ____	6 + 5 = ____	6 + 9 = ____
6 + 0 = ____	6 + 0 = ____	6 + 5 = ____	6 + 1 = ____	6 + 2 = ____
6 + 5 = ____	6 + 6 = ____	6 + 3 = ____	6 + 8 = ____	6 + 6 = ____
6 + 2 = ____	6 + 9 = ____	6 + 8 = ____	6 + 0 = ____	6 + 4 = ____
6 + 7 = ____	6 + 1 = ____	6 + 4 = ____	6 + 6 = ____	6 + 1 = ____
6 + 9 = ____	6 + 3 = ____	6 + 0 = ____	6 + 2 = ____	6 + 5 = ____
6 + 3 = ____	6 + 5 = ____	6 + 1 = ____	6 + 4 = ____	6 + 0 = ____
6 + 8 = ____	6 + 8 = ____	6 + 6 = ____	6 + 7 = ____	6 + 3 = ____
6 + 1 = ____	6 + 2 = ____	6 + 9 = ____	6 + 3 = ____	6 + 8 = ____
6 + 6 = ____	6 + 4 = ____	6 + 2 = ____	6 + 5 = ____	6 + 7 = ____
6 + 4 = ____	6 + 9 = ____	6 + 7 = ____	6 + 1 = ____	6 + 9 = ____
Score: ____ /25 _____ Min. _____ Sec.	Score: ____ /25 _____ Min. _____ Sec.	Score: ____ /25 _____ Min. _____ Sec.	Score: ____ /25 _____ Min. _____ Sec.	Score: ____ /25 _____ Min. _____ Sec.

OTM-1139 • SSK1-39 Timed Addition Facts

Extra Practice Six Plus Drills

Name: _____

Day 1	Day 2	Day 3	Day 4	Day 5
6 + ___ = 9	___ + 2 = 8	6 + ___ = 12	___ + 1 = 7	6 + ___ = 10
___ + 5 = 11	6 + ___ = 10	___ + 4 = 12	6 + ___ = 12	___ + 0 = 6
6 + ___ = 14	___ + 7 = 13	6 + ___ = 6	___ + 9 = 15	6 + ___ = 14
___ + 2 = 8	6 + ___ = 9	___ + 5 = 11	6 + ___ = 8	___ + 7 = 13
6 + ___ = 10	___ + 5 = 11	6 + ___ = 8	___ + 7 = 13	6 + ___ = 15
___ + 0 = 6	6 + ___ = 7	___ + 7 = 13	6 + ___ = 11	___ + 2 = 8
6 + ___ = 12	___ + 8 = 14	6 + ___ = 15	___ + 3 = 9	6 + ___ = 12
___ + 9 = 15	6 + ___ = 6	___ + 3 = 9	6 + ___ = 14	___ + 4 = 10
6 + ___ = 13	___ + 9 = 15	6 + ___ = 14	___ + 4 = 10	6 + ___ = 7
___ + 1 = 7	6 + ___ = 12	___ + 1 = 7	6 + ___ = 6	___ + 5 = 11
6 + ___ = 9	___ + 2 = 8	6 + ___ = 12	___ + 1 = 7	6 + ___ = 6
___ + 5 = 11	6 + ___ = 10	___ + 4 = 10	6 + ___ = 12	___ + 3 = 9
6 + ___ = 14	___ + 7 = 13	6 + ___ = 6	___ + 9 = 15	6 + ___ = 14
___ + 2 = 8	6 + ___ = 9	___ + 2 = 8	6 + ___ = 8	___ + 5 = 11
6 + ___ = 10	___ + 5 = 11	6 + ___ = 10	___ + 7 = 13	6 + ___ = 15
___ + 0 = 6	6 + ___ = 7	___ + 0 = 11	6 + ___ = 11	___ + 2 = 8
6 + ___ = 12	___ + 8 = 14	6 + ___ = 11	___ + 3 = 9	6 + ___ = 12
___ + 9 = 15	6 + ___ = 6	___ + 2 = 8	6 + ___ = 14	___ + 4 = 10
6 + ___ = 7	___ + 6 = 12	6 + ___ = 13	___ + 4 = 10	6 + ___ = 7
___ + 3 = 9	6 + ___ = 8	___ + 9 = 15	6 + ___ = 6	___ + 5 = 11
6 + ___ = 11	___ + 4 = 10	6 + ___ = 9	___ + 1 = 7	6 + ___ = 6
___ + 8 = 14	6 + ___ = 13	___ + 8 = 13	6 + ___ = 12	___ + 3 = 9
6 + ___ = 8	___ + 3 = 9	6 + ___ = 7	___ + 9 = 15	6 + ___ = 14
___ + 4 = 10	6 + ___ = 11	___ + 6 = 12	6 + ___ = 8	___ + 7 = 13
6 + ___ = 15	___ + 1 = 7	6 + ___ = 10	___ + 7 = 13	6 + ___ = 15
Score: ____/25	Score: ____/25	Score: ____/25	Score: ____/25	Score: ____/25
_____ Min.	_____ Min.	_____ Min.	_____ Min.	_____ Min.
_____ Sec.	_____ Sec.	_____ Sec.	_____ Sec.	_____ Sec.

Six Plus Review Test

Name: _____

6 +3	6 +5	6 +8	6 +2	6 +4	6 +0	6 +6	6 +9	6 +7	6 +1
6 +8	6 +3	6 +2	6 +5	6 +9	6 +1	6 +4	6 +3	6 +0	6 +6
6 +5	6 +8	6 +2	6 +4	6 +7	6 +3	6 +5	6 +9	6 +1	6 +8
6 +0	6 +9	6 +6	6 +7	6 +2	6 +4	6 +3	6 +5	6 +8	6 +4
6 +1	6 +0	6 +2	6 +4	6 +7	6 +3	6 +5	6 +1	6 +6	6 +0
6 +5	6 +2	6 +4	6 +6	6 +2	6 +7	6 +9	6 +3	6 +8	6 +1
6 +6	6 +4	6 +0	6 +2	6 +4	6 +0	6 +5	6 +2	6 +7	6 +9
6 +3	6 +8	6 +1	6 +6	6 +4	6 +1	6 +6	6 +9	6 +2	6 +7
6 +5	6 +3	6 +8	6 +4	6 +0	6 +1	6 +9	6 +7	6 +5	6 +3
6 +8	6 +4	6 +0	6 +1	6 +6	6 +9	6 +2	6 +7	6 +4	6 +8

Date: _____ Score: _____/100 Time: _____ Min. _____ Sec.

 OTM-1139 • SSK1-39 Timed Addition Facts

Seven Plus Drills

Name: _____

Date: Monday _____ Score: _____ /25 Time: _____ Min. _____ Sec.

7 + 3 = ___	7 + 2 = ___	7 + 6 = ___	7 + 1 = ___	7 + 4 = ___
7 + 5 = ___	7 + 4 = ___	7 + 4 = ___	7 + 6 = ___	7 + 0 = ___
7 + 8 = ___	7 + 7 = ___	7 + 0 = ___	7 + 9 = ___	7 + 8 = ___
7 + 2 = ___	7 + 3 = ___	7 + 5 = ___	7 + 2 = ___	7 + 7 = ___
7 + 4 = ___	7 + 5 = ___	7 + 2 = ___	7 + 7 = ___	7 + 9 = ___

Date: Tuesday _____ Score: _____ /25 Time: _____ Min. _____ Sec.

7 + 0 = ___	7 + 1 = ___	7 + 7 = ___	7 + 5 = ___	7 + 2 = ___
7 + 6 = ___	7 + 8 = ___	7 + 9 = ___	7 + 3 = ___	7 + 6 = ___
7 + 9 = ___	7 + 0 = ___	7 + 3 = ___	7 + 8 = ___	7 + 4 = ___
7 + 7 = ___	7 + 9 = ___	7 + 8 = ___	7 + 4 = ___	7 + 1 = ___
7 + 1 = ___	7 + 6 = ___	7 + 1 = ___	7 + 0 = ___	7 + 5 = ___

Date: Wednesday _____ Score: _____ /25 Time: _____ Min. _____ Sec.

7 + 3 = ___	7 + 2 = ___	7 + 6 = ___	7 + 1 = ___	7 + 0 = ___
7 + 5 = ___	7 + 4 = ___	7 + 4 = ___	7 + 6 = ___	7 + 3 = ___
7 + 8 = ___	7 + 7 = ___	7 + 0 = ___	7 + 9 = ___	7 + 8 = ___
7 + 2 = ___	7 + 3 = ___	7 + 2 = ___	7 + 2 = ___	7 + 5 = ___
7 + 4 = ___	7 + 5 = ___	7 + 4 = ___	7 + 7 = ___	7 + 9 = ___

Date: Thursday _____ Score: _____ /25 Time: _____ Min. _____ Sec.

7 + 0 = ___	7 + 1 = ___	7 + 0 = ___	7 + 5 = ___	7 + 2 = ___
7 + 6 = ___	7 + 8 = ___	7 + 5 = ___	7 + 3 = ___	7 + 6 = ___
7 + 9 = ___	7 + 0 = ___	7 + 2 = ___	7 + 8 = ___	7 + 4 = ___
7 + 1 = ___	7 + 6 = ___	7 + 7 = ___	7 + 4 = ___	7 + 1 = ___
7 + 3 = ___	7 + 2 = ___	7 + 9 = ___	7 + 0 = ___	7 + 5 = ___

Date: Friday _____ Score: _____ /25 Time: _____ Min. _____ Sec.

7 + 5 = ___	7 + 4 = ___	7 + 3 = ___	7 + 1 = ___	7 + 0 = ___
7 + 8 = ___	7 + 7 = ___	7 + 8 = ___	7 + 6 = ___	7 + 3 = ___
7 + 2 = ___	7 + 3 = ___	7 + 1 = ___	7 + 9 = ___	7 + 8 = ___
7 + 4 = ___	7 + 5 = ___	7 + 6 = ___	7 + 2 = ___	7 + 7 = ___
7 + 9 = ___	7 + 1 = ___	7 + 4 = ___	7 + 7 = ___	7 + 9 = ___

Home Practice Seven Plus Drills

Name: _____

Monday	Tuesday	Wednesday	Thursday	Friday
7 + 3 = ___	7 + 1 = ___	7 + 4 = ___	7 + 2 = ___	7 + 6 = ___
7 + 5 = ___	7 + 6 = ___	7 + 0 = ___	7 + 4 = ___	7 + 4 = ___
7 + 8 = ___	7 + 9 = ___	7 + 8 = ___	7 + 7 = ___	7 + 0 = ___
7 + 2 = ___	7 + 2 = ___	7 + 7 = ___	7 + 3 = ___	7 + 5 = ___
7 + 4 = ___	7 + 7 = ___	7 + 9 = ___	7 + 5 = ___	7 + 2 = ___
7 + 0 = ___	7 + 5 = ___	7 + 2 = ___	7 + 1 = ___	7 + 7 = ___
7 + 6 = ___	7 + 3 = ___	7 + 6 = ___	7 + 8 = ___	7 + 9 = ___
7 + 9 = ___	7 + 8 = ___	7 + 4 = ___	7 + 0 = ___	7 + 3 = ___
7 + 7 = ___	7 + 4 = ___	7 + 1 = ___	7 + 9 = ___	7 + 8 = ___
7 + 1 = ___	7 + 0 = ___	7 + 5 = ___	7 + 6 = ___	7 + 1 = ___
7 + 3 = ___	7 + 1 = ___	7 + 0 = ___	7 + 2 = ___	7 + 6 = ___
7 + 5 = ___	7 + 6 = ___	7 + 3 = ___	7 + 4 = ___	7 + 4 = ___
7 + 8 = ___	7 + 9 = ___	7 + 8 = ___	7 + 7 = ___	7 + 0 = ___
7 + 2 = ___	7 + 2 = ___	7 + 5 = ___	7 + 3 = ___	7 + 2 = ___
7 + 4 = ___	7 + 7 = ___	7 + 9 = ___	7 + 5 = ___	7 + 4 = ___
7 + 0 = ___	7 + 5 = ___	7 + 2 = ___	7 + 1 = ___	7 + 0 = ___
7 + 6 = ___	7 + 3 = ___	7 + 6 = ___	7 + 8 = ___	7 + 5 = ___
7 + 9 = ___	7 + 8 = ___	7 + 4 = ___	7 + 0 = ___	7 + 2 = ___
7 + 1 = ___	7 + 4 = ___	7 + 1 = ___	7 + 6 = ___	7 + 7 = ___
7 + 3 = ___	7 + 0 = ___	7 + 5 = ___	7 + 2 = ___	7 + 9 = ___
7 + 5 = ___	7 + 1 = ___	7 + 0 = ___	7 + 4 = ___	7 + 3 = ___
7 + 8 = ___	7 + 6 = ___	7 + 3 = ___	7 + 7 = ___	7 + 8 = ___
7 + 2 = ___	7 + 9 = ___	7 + 8 = ___	7 + 3 = ___	7 + 1 = ___
7 + 4 = ___	7 + 2 = ___	7 + 7 = ___	7 + 5 = ___	7 + 6 = ___
7 + 9 = ___	7 + 7 = ___	7 + 9 = ___	7 + 1 = ___	7 + 4 = ___
Score: ____/25 _____ Min. _____ Sec.	Score: ____/25 _____ Min. _____ Sec.	Score: ____/25 _____ Min. _____ Sec.	Score: ____/25 _____ Min. _____ Sec.	Score: ____/25 _____ Min. _____ Sec.

Extra Practice Seven Plus Drills

Name: _____

Day 1	Day 2	Day 3	Day 4	Day 5
7 + ___ = 10	___ + 8 = 15	7 + ___ = 13	___ + 1 = 8	7 + ___ = 11
___ + 5 = 12	7 + ___ = 9	___ + 4 = 11	7 + ___ = 13	___ + 0 = 7
7 + ___ = 15	___ + 4 = 11	7 + ___ = 7	___ + 9 = 16	7 + ___ = 15
___ + 2 = 9	7 + ___ = 16	___ + 5 = 12	7 + ___ = 9	___ + 7 = 14
7 + ___ = 11	___ + 2 = 9	7 + ___ = 9	___ + 7 = 14	7 + ___ = 16
___ + 0 = 7	7 + ___ = 11	___ + 7 = 14	7 + ___ = 12	___ + 2 = 9
7 + ___ = 13	___ + 7 = 14	7 + ___ = 16	___ + 3 = 10	7 + ___ = 13
___ + 9 = 16	7 + ___ = 10	___ + 3 = 10	7 + ___ = 15	___ + 4 = 11
7 + ___ = 14	___ + 5 = 12	7 + ___ = 15	___ + 4 = 11	7 + ___ = 8
___ + 1 = 8	7 + ___ = 8	___ + 1 = 8	7 + ___ = 7	___ + 5 = 12
7 + ___ = 10	___ + 8 = 15	7 + ___ = 13	___ + 1 = 8	7 + ___ = 7
___ + 5 = 12	7 + ___ = 7	___ + 4 = 11	7 + ___ = 13	___ + 3 = 10
7 + ___ = 15	___ + 9 = 16	7 + ___ = 7	___ + 9 = 16	7 + ___ = 15
___ + 2 = 9	7 + ___ = 13	___ + 2 = 9	7 + ___ = 9	___ + 5 = 12
7 + ___ = 11	___ + 2 = 9	7 + ___ = 14	___ + 7 = 14	7 + ___ = 16
___ + 0 = 7	7 + ___ = 11	___ + 0 = 7	7 + ___ = 12	___ + 2 = 9
7 + ___ = 13	___ + 7 = 14	7 + ___ = 12	___ + 3 = 10	7 + ___ = 13
___ + 9 = 16	7 + ___ = 10	___ + 2 = 9	7 + ___ = 15	___ + 4 = 11
7 + ___ = 8	___ + 5 = 12	7 + ___ = 14	___ + 4 = 11	7 + ___ = 8
___ + 3 = 10	7 + ___ = 8	___ + 9 = 16	7 + ___ = 7	___ + 5 = 12
7 + ___ = 13	___ + 0 = 7	7 + ___ = 10	___ + 1 = 8	7 + ___ = 7
___ + 9 = 16	___ + 6 = 13	___ + 8 = 15	7 + ___ = 13	___ + 3 = 10
7 + ___ = 8	7 + ___ = 9	7 + ___ = 8	___ + 9 = 16	7 + ___ = 15
___ + 3 = 10	___ + 4 = 11	___ + 6 = 13	7 + ___ = 9	___ + 7 = 14
7 + ___ = 12	7 + ___ = 14	7 + ___ = 11	___ + ___ = 14	7 + ___ = 16
Score: _____/25	Score: _____/25	Score: _____/25	Score: _____/25	Score: _____/25
_____ Min.	_____ Min.	_____ Min.	_____ Min.	_____ Min.
_____ Sec.	_____ Sec.	_____ Sec.	_____ Sec.	_____ Sec.

Seven Plus Review Test

Name: _____

7 +3	7 +5	7 +8	7 +2	7 +4	7 +0	7 +6	7 +9	7 +7	7 +1
7 +0	7 +4	7 +3	7 +6	7 +8	7 +5	7 +1	7 +2	7 +8	7 +9
7 +2	7 +8	7 +1	7 +7	7 +2	7 +4	7 +0	7 +5	7 +9	7 +3
7 +6	7 +2	7 +4	7 +9	7 +4	7 +3	7 +7	7 +3	7 +5	7 +1
7 +8	7 +0	7 +6	7 +2	7 +7	7 +5	7 +4	7 +1	7 +6	7 +3
7 +4	7 +2	7 +5	7 +3	7 +8	7 +0	7 +1	7 +7	7 +6	7 +9
7 +0	7 +4	7 +2	7 +4	7 +9	7 +7	7 +0	7 +3	7 +5	7 +2
7 +8	7 +2	7 +1	7 +9	7 +6	7 +5	7 +4	7 +7	7 +1	7 +6
7 +3	7 +4	7 +6	7 +7	7 +2	7 +3	7 +0	7 +4	7 +6	7 +9
7 +8	7 +0	7 +1	7 +5	7 +9	7 +8	7 +2	7 +7	7 +4	7 +1

Date: _____ Score: _____/100 Time: _____ Min. _____ Sec.

Eight Plus Drills

Name: _____

Date: Monday _____ Score: _____ /25 Time: _____ Min. _____ Sec.

8 + 5 = ___	8 + 4 = ___	8 + 3 = ___	8 + 1 = ___	8 + 0 = ___
8 + 8 = ___	8 + 7 = ___	8 + 8 = ___	8 + 6 = ___	8 + 3 = ___
8 + 2 = ___	8 + 3 = ___	8 + 4 = ___	8 + 9 = ___	8 + 8 = ___
8 + 4 = ___	8 + 5 = ___	8 + 6 = ___	8 + 2 = ___	8 + 7 = ___
8 + 9 = ___	8 + 1 = ___	8 + 4 = ___	8 + 7 = ___	8 + 9 = ___

Date: Tuesday _____ Score: _____ /25 Time: _____ Min. _____ Sec.

8 + 0 = ___	8 + 1 = ___	8 + 0 = ___	8 + 5 = ___	8 + 2 = ___
8 + 6 = ___	8 + 8 = ___	8 + 5 = ___	8 + 3 = ___	8 + 6 = ___
8 + 9 = ___	8 + 0 = ___	8 + 2 = ___	8 + 8 = ___	8 + 4 = ___
8 + 1 = ___	8 + 6 = ___	8 + 7 = ___	8 + 4 = ___	8 + 1 = ___
8 + 3 = ___	8 + 2 = ___	8 + 9 = ___	8 + 0 = ___	8 + 5 = ___

Date: Wednesday _____ Score: _____ /25 Time: _____ Min. _____ Sec.

8 + 3 = ___	8 + 2 = ___	8 + 6 = ___	8 + 1 = ___	8 + 4 = ___
8 + 5 = ___	8 + 4 = ___	8 + 4 = ___	8 + 6 = ___	8 + 0 = ___
8 + 8 = ___	8 + 7 = ___	8 + 0 = ___	8 + 9 = ___	8 + 8 = ___
8 + 2 = ___	8 + 3 = ___	8 + 5 = ___	8 + 2 = ___	8 + 7 = ___
8 + 4 = ___	8 + 5 = ___	8 + 2 = ___	8 + 7 = ___	8 + 9 = ___

Date: Thursday _____ Score: _____ /25 Time: _____ Min. _____ Sec.

8 + 0 = ___	8 + 1 = ___	8 + 7 = ___	8 + 5 = ___	8 + 2 = ___
8 + 6 = ___	8 + 8 = ___	8 + 9 = ___	8 + 3 = ___	8 + 6 = ___
8 + 9 = ___	8 + 0 = ___	8 + 3 = ___	8 + 8 = ___	8 + 4 = ___
8 + 7 = ___	8 + 9 = ___	8 + 8 = ___	8 + 4 = ___	8 + 1 = ___
8 + 1 = ___	8 + 6 = ___	8 + 1 = ___	8 + 0 = ___	8 + 5 = ___

Date: Friday _____ Score: _____ /25 Time: _____ Min. _____ Sec.

8 + 3 = ___	8 + 2 = ___	8 + 6 = ___	8 + 1 = ___	8 + 0 = ___
8 + 5 = ___	8 + 4 = ___	8 + 4 = ___	8 + 6 = ___	8 + 3 = ___
8 + 8 = ___	8 + 7 = ___	8 + 0 = ___	8 + 9 = ___	8 + 8 = ___
8 + 2 = ___	8 + 3 = ___	8 + 2 = ___	8 + 2 = ___	8 + 5 = ___
8 + 4 = ___	8 + 5 = ___	8 + 4 = ___	8 + 7 = ___	8 + 9 = ___

Home Practice Eight Plus Drills

Name: _____

Monday	Tuesday	Wednesday	Thursday	Friday
8 + 1 = ____	8 + 4 = ____	8 + 3 = ____	8 + 2 = ____	8 + 6 = ____
8 + 6 = ____	8 + 0 = ____	8 + 5 = ____	8 + 4 = ____	8 + 4 = ____
8 + 9 = ____	8 + 8 = ____	8 + 8 = ____	8 + 7 = ____	8 + 0 = ____
8 + 2 = ____	8 + 7 = ____	8 + 2 = ____	8 + 3 = ____	8 + 5 = ____
8 + 7 = ____	8 + 9 = ____	8 + 4 = ____	8 + 5 = ____	8 + 2 = ____
8 + 5 = ____	8 + 2 = ____	8 + 0 = ____	8 + 1 = ____	8 + 7 = ____
8 + 3 = ____	8 + 6 = ____	8 + 6 = ____	8 + 8 = ____	8 + 9 = ____
8 + 8 = ____	8 + 4 = ____	8 + 9 = ____	8 + 0 = ____	8 + 3 = ____
8 + 4 = ____	8 + 1 = ____	8 + 7 = ____	8 + 9 = ____	8 + 8 = ____
8 + 0 = ____	8 + 5 = ____	8 + 1 = ____	8 + 6 = ____	8 + 1 = ____
8 + 1 = ____	8 + 0 = ____	8 + 3 = ____	8 + 2 = ____	8 + 6 = ____
8 + 6 = ____	8 + 3 = ____	8 + 5 = ____	8 + 4 = ____	8 + 4 = ____
8 + 9 = ____	8 + 8 = ____	8 + 8 = ____	8 + 7 = ____	8 + 0 = ____
8 + 2 = ____	8 + 5 = ____	8 + 2 = ____	8 + 3 = ____	8 + 2 = ____
8 + 7 = ____	8 + 9 = ____	8 + 4 = ____	8 + 5 = ____	8 + 4 = ____
8 + 5 = ____	8 + 2 = ____	8 + 0 = ____	8 + 1 = ____	8 + 0 = ____
8 + 3 = ____	8 + 6 = ____	8 + 6 = ____	8 + 8 = ____	8 + 5 = ____
8 + 8 = ____	8 + 4 = ____	8 + 9 = ____	8 + 0 = ____	8 + 2 = ____
8 + 4 = ____	8 + 1 = ____	8 + 1 = ____	8 + 6 = ____	8 + 7 = ____
8 + 0 = ____	8 + 5 = ____	8 + 3 = ____	8 + 2 = ____	8 + 9 = ____
8 + 1 = ____	8 + 0 = ____	8 + 5 = ____	8 + 4 = ____	8 + 3 = ____
8 + 6 = ____	8 + 3 = ____	8 + 8 = ____	8 + 7 = ____	8 + 8 = ____
8 + 9 = ____	8 + 8 = ____	8 + 2 = ____	8 + 3 = ____	8 + 1 = ____
8 + 2 = ____	8 + 7 = ____	8 + 4 = ____	8 + 5 = ____	8 + 6 = ____
8 + 7 = ____	8 + 9 = ____	7 + 9 = ____	8 + 1 = ____	8 + 4 = ____
Score: ____/25 ____ Min. ____ Sec.	Score: ____/25 ____ Min. ____ Sec.	Score: ____/25 ____ Min. ____ Sec.	Score: ____/25 ____ Min. ____ Sec.	Score: ____/25 ____ Min. ____ Sec.

Extra Practice Eight Plus Drills

Name: _____

Day 1	Day 2	Day 3	Day 4	Day 5
$8 + \underline{\hphantom{0}} = 11$	$\underline{\hphantom{0}} + 2 = 10$	$8 + \underline{\hphantom{0}} = 14$	$\underline{\hphantom{0}} + 1 = 9$	$8 + \underline{\hphantom{0}} = 12$
$\underline{\hphantom{0}} + 5 = 13$	$8 + \underline{\hphantom{0}} = 12$	$\underline{\hphantom{0}} + 4 = 12$	$8 + \underline{\hphantom{0}} = 14$	$\underline{\hphantom{0}} + 0 = 8$
$8 + \underline{\hphantom{0}} = 16$	$\underline{\hphantom{0}} + 7 = 15$	$8 + \underline{\hphantom{0}} = 8$	$\underline{\hphantom{0}} + 9 = 17$	$8 + \underline{\hphantom{0}} = 16$
$\underline{\hphantom{0}} + 2 = 10$	$8 + \underline{\hphantom{0}} = 11$	$\underline{\hphantom{0}} + 5 = 13$	$8 + \underline{\hphantom{0}} = 10$	$\underline{\hphantom{0}} + 7 = 15$
$8 + \underline{\hphantom{0}} = 12$	$\underline{\hphantom{0}} + 5 = 13$	$8 + \underline{\hphantom{0}} = 10$	$\underline{\hphantom{0}} + 7 = 15$	$8 + \underline{\hphantom{0}} = 17$
$\underline{\hphantom{0}} + 0 = 8$	$8 + \underline{\hphantom{0}} = 9$	$\underline{\hphantom{0}} + 7 = 15$	$8 + \underline{\hphantom{0}} = 13$	$\underline{\hphantom{0}} + 2 = 10$
$8 + \underline{\hphantom{0}} = 14$	$\underline{\hphantom{0}} + 8 = 16$	$8 + \underline{\hphantom{0}} = 17$	$\underline{\hphantom{0}} + 3 = 11$	$8 + \underline{\hphantom{0}} = 14$
$\underline{\hphantom{0}} + 9 = 17$	$8 + \underline{\hphantom{0}} = 8$	$\underline{\hphantom{0}} + 3 = 11$	$8 + \underline{\hphantom{0}} = 16$	$\underline{\hphantom{0}} + 4 = 12$
$8 + \underline{\hphantom{0}} = 16$	$\underline{\hphantom{0}} + 9 = 17$	$8 + \underline{\hphantom{0}} = 16$	$\underline{\hphantom{0}} + 4 = 12$	$8 + \underline{\hphantom{0}} = 9$
$\underline{\hphantom{0}} + 1 = 9$	$8 + \underline{\hphantom{0}} = 14$	$\underline{\hphantom{0}} + 1 = 9$	$8 + \underline{\hphantom{0}} = 8$	$\underline{\hphantom{0}} + 5 = 13$
$8 + \underline{\hphantom{0}} = 11$	$\underline{\hphantom{0}} + 2 = 10$	$8 + \underline{\hphantom{0}} = 14$	$\underline{\hphantom{0}} + 1 = 9$	$8 + \underline{\hphantom{0}} = 8$
$\underline{\hphantom{0}} + 5 = 13$	$8 + \underline{\hphantom{0}} = 12$	$\underline{\hphantom{0}} + 4 = 12$	$8 + \underline{\hphantom{0}} = 14$	$\underline{\hphantom{0}} + 3 = 11$
$8 + \underline{\hphantom{0}} = 16$	$\underline{\hphantom{0}} + 7 = 15$	$8 + \underline{\hphantom{0}} = 8$	$\underline{\hphantom{0}} + 9 = 17$	$8 + \underline{\hphantom{0}} = 16$
$\underline{\hphantom{0}} + 2 = 10$	$8 + \underline{\hphantom{0}} = 11$	$\underline{\hphantom{0}} + 2 = 10$	$8 + \underline{\hphantom{0}} = 10$	$\underline{\hphantom{0}} + 5 = 13$
$8 + \underline{\hphantom{0}} = 12$	$\underline{\hphantom{0}} + 5 = 13$	$8 + \underline{\hphantom{0}} = 12$	$\underline{\hphantom{0}} + 7 = 15$	$8 + \underline{\hphantom{0}} = 17$
$\underline{\hphantom{0}} + 0 = 8$	$8 + \underline{\hphantom{0}} = 9$	$\underline{\hphantom{0}} + 0 = 8$	$8 + \underline{\hphantom{0}} = 13$	$\underline{\hphantom{0}} + 2 = 10$
$8 + \underline{\hphantom{0}} = 14$	$\underline{\hphantom{0}} + 8 = 16$	$8 + \underline{\hphantom{0}} = 13$	$\underline{\hphantom{0}} + 3 = 11$	$8 + \underline{\hphantom{0}} = 14$
$\underline{\hphantom{0}} + 9 = 17$	$8 + \underline{\hphantom{0}} = 8$	$\underline{\hphantom{0}} + 2 = 10$	$8 + \underline{\hphantom{0}} = 16$	$\underline{\hphantom{0}} + 4 = 12$
$8 + \underline{\hphantom{0}} = 9$	$\underline{\hphantom{0}} + 6 = 14$	$8 + \underline{\hphantom{0}} = 15$	$\underline{\hphantom{0}} + 4 = 12$	$8 + \underline{\hphantom{0}} = 9$
$\underline{\hphantom{0}} + 3 = 11$	$8 + \underline{\hphantom{0}} = 10$	$\underline{\hphantom{0}} + 9 = 17$	$8 + \underline{\hphantom{0}} = 8$	$\underline{\hphantom{0}} + 5 = 13$
$8 + \underline{\hphantom{0}} = 13$	$\underline{\hphantom{0}} + 4 = 12$	$8 + \underline{\hphantom{0}} = 11$	$\underline{\hphantom{0}} + 1 = 9$	$8 + \underline{\hphantom{0}} = 8$
$\underline{\hphantom{0}} + 8 = 16$	$8 + \underline{\hphantom{0}} = 15$	$\underline{\hphantom{0}} + 8 = 16$	$8 + \underline{\hphantom{0}} = 14$	$\underline{\hphantom{0}} + 3 = 11$
$8 + \underline{\hphantom{0}} = 10$	$\underline{\hphantom{0}} + 3 = 11$	$8 + \underline{\hphantom{0}} = 9$	$\underline{\hphantom{0}} + 9 = 17$	$8 + \underline{\hphantom{0}} = 16$
$\underline{\hphantom{0}} + 4 = 12$	$8 + \underline{\hphantom{0}} = 13$	$\underline{\hphantom{0}} + 6 = 14$	$8 + \underline{\hphantom{0}} = 10$	$\underline{\hphantom{0}} + 7 = 15$
$8 + \underline{\hphantom{0}} = 17$	$\underline{\hphantom{0}} + 1 = 9$	$8 + \underline{\hphantom{0}} = 12$	$\underline{\hphantom{0}} + 7 = 15$	$8 + \underline{\hphantom{0}} = 17$
Score: _____ /25	Score: _____ /25	Score: _____ /25	Score: _____ /25	Score: _____ /25
_____ Min.	_____ Min.	_____ Min.	_____ Min.	_____ Min.
_____ Sec.	_____ Sec.	_____ Sec.	_____ Sec.	_____ Sec.

Eight Plus Review Test

Name: _____

8 +3	8 +5	8 +8	8 +2	8 +4	8 +0	8 +6	8 +9	8 +7	8 +1
8 +2	8 +3	8 +4	8 +0	8 +5	8 +9	8 +8	8 +1	8 +3	8 +6
8 +5	8 +8	8 +2	8 +4	8 +2	8 +3	8 +7	8 +5	8 +4	8 +9
8 +1	8 +9	8 +7	8 +0	8 +4	8 +8	8 +6	8 +3	8 +2	8 +5
8 +8	8 +4	8 +6	8 +1	8 +7	8 +0	8 +3	8 +6	8 +5	8 +2
8 +4	8 +7	8 +1	8 +3	8 +5	8 +8	8 +2	8 +1	8 +9	8 +0
8 +6	8 +4	8 +0	8 +2	8 +4	8 +0	8 +5	8 +2	8 +7	8 +9
8 +3	8 +8	8 +1	8 +6	8 +9	8 +2	8 +4	8 +5	8 +3	8 +1
8 +8	8 +4	8 +0	8 +1	8 +7	8 +6	8 +9	8 +2	8 +6	8 +7
8 +5	8 +3	8 +8	8 +4	8 +0	8 +1	8 +6	8 +9	8 +2	8 +7

Date: _____   Score: _____/100   Time: _____ Min. _____ Sec.

Nine Plus Drills

Name: _____

Date: Monday _____ Score: _____ /25 Time: _____ Min. _____ Sec.

9 + 0 = ___	9 + 1 = ___	9 + 0 = ___	9 + 5 = ___	9 + 2 = ___
9 + 6 = ___	9 + 8 = ___	9 + 5 = ___	9 + 3 = ___	9 + 6 = ___
9 + 9 = ___	9 + 0 = ___	9 + 2 = ___	9 + 8 = ___	9 + 4 = ___
9 + 1 = ___	9 + 6 = ___	9 + 7 = ___	9 + 4 = ___	9 + 1 = ___
9 + 3 = ___	9 + 2 = ___	9 + 9 = ___	9 + 0 = ___	9 + 5 = ___

Date: Tuesday _____ Score: _____ /25 Time: _____ Min. _____ Sec.

9 + 5 = ___	9 + 4 = ___	9 + 3 = ___	9 + 1 = ___	9 + 0 = ___
9 + 8 = ___	9 + 7 = ___	9 + 8 = ___	9 + 6 = ___	9 + 3 = ___
9 + 2 = ___	9 + 3 = ___	9 + 1 = ___	9 + 9 = ___	9 + 8 = ___
9 + 4 = ___	9 + 5 = ___	9 + 6 = ___	9 + 2 = ___	9 + 7 = ___
9 + 9 = ___	9 + 1 = ___	9 + 4 = ___	9 + 7 = ___	9 + 9 = ___

Date: Wednesday _____ Score: _____ /25 Time: _____ Min. _____ Sec.

9 + 0 = ___	9 + 1 = ___	9 + 7 = ___	9 + 5 = ___	9 + 2 = ___
9 + 6 = ___	9 + 8 = ___	9 + 9 = ___	9 + 3 = ___	9 + 6 = ___
9 + 9 = ___	9 + 0 = ___	9 + 3 = ___	9 + 8 = ___	9 + 4 = ___
9 + 7 = ___	9 + 9 = ___	9 + 8 = ___	9 + 4 = ___	9 + 1 = ___
9 + 1 = ___	9 + 6 = ___	9 + 1 = ___	9 + 0 = ___	9 + 5 = ___

Date: Thursday _____ Score: _____ /25 Time: _____ Min. _____ Sec.

9 + 3 = ___	9 + 2 = ___	9 + 6 = ___	9 + 1 = ___	9 + 0 = ___
9 + 5 = ___	9 + 4 = ___	9 + 4 = ___	9 + 6 = ___	9 + 3 = ___
9 + 8 = ___	9 + 7 = ___	9 + 0 = ___	9 + 9 = ___	9 + 8 = ___
9 + 2 = ___	9 + 3 = ___	9 + 2 = ___	9 + 2 = ___	9 + 5 = ___
9 + 4 = ___	9 + 5 = ___	9 + 4 = ___	9 + 7 = ___	9 + 9 = ___

Date: Friday _____ Score: _____ /25 Time: _____ Min. _____ Sec.

9 + 3 = ___	9 + 2 = ___	9 + 6 = ___	9 + 1 = ___	9 + 4 = ___
9 + 5 = ___	9 + 4 = ___	9 + 4 = ___	9 + 6 = ___	9 + 0 = ___
9 + 8 = ___	9 + 7 = ___	9 + 0 = ___	9 + 9 = ___	9 + 8 = ___
9 + 2 = ___	9 + 3 = ___	9 + 5 = ___	9 + 2 = ___	9 + 7 = ___
9 + 4 = ___	9 + 5 = ___	9 + 2 = ___	9 + 7 = ___	9 + 9 = ___

Home Practice Nine Plus Drills

Name: _____

Monday	Tuesday	Wednesday	Thursday	Friday
9 + 3 = ___	9 + 6 = ___	9 + 4 = ___	9 + 1 = ___	9 + 2 = ___
9 + 5 = ___	9 + 4 = ___	9 + 0 = ___	9 + 6 = ___	9 + 4 = ___
9 + 8 = ___	9 + 0 = ___	9 + 8 = ___	9 + 9 = ___	9 + 7 = ___
9 + 2 = ___	9 + 5 = ___	9 + 7 = ___	9 + 2 = ___	9 + 3 = ___
9 + 4 = ___	9 + 2 = ___	9 + 9 = ___	9 + 7 = ___	9 + 5 = ___
9 + 0 = ___	9 + 7 = ___	9 + 2 = ___	9 + 5 = ___	9 + 1 = ___
9 + 6 = ___	9 + 9 = ___	9 + 6 = ___	9 + 3 = ___	9 + 8 = ___
9 + 9 = ___	9 + 3 = ___	9 + 4 = ___	9 + 8 = ___	9 + 0 = ___
9 + 7 = ___	9 + 8 = ___	9 + 1 = ___	9 + 4 = ___	9 + 9 = ___
9 + 1 = ___	9 + 1 = ___	9 + 5 = ___	9 + 0 = ___	9 + 6 = ___
9 + 3 = ___	9 + 6 = ___	9 + 0 = ___	9 + 1 = ___	9 + 2 = ___
9 + 5 = ___	9 + 4 = ___	9 + 3 = ___	9 + 6 = ___	9 + 4 = ___
9 + 8 = ___	9 + 0 = ___	9 + 8 = ___	9 + 9 = ___	9 + 7 = ___
9 + 2 = ___	9 + 2 = ___	9 + 5 = ___	9 + 2 = ___	9 + 3 = ___
9 + 4 = ___	9 + 4 = ___	9 + 9 = ___	9 + 7 = ___	9 + 5 = ___
9 + 0 = ___	9 + 0 = ___	9 + 2 = ___	9 + 5 = ___	9 + 1 = ___
9 + 6 = ___	9 + 5 = ___	9 + 6 = ___	9 + 3 = ___	9 + 8 = ___
9 + 9 = ___	9 + 2 = ___	9 + 4 = ___	9 + 8 = ___	9 + 0 = ___
9 + 1 = ___	9 + 7 = ___	9 + 1 = ___	9 + 4 = ___	9 + 6 = ___
9 + 3 = ___	9 + 9 = ___	9 + 5 = ___	9 + 0 = ___	9 + 2 = ___
9 + 5 = ___	9 + 3 = ___	9 + 0 = ___	9 + 1 = ___	9 + 4 = ___
9 + 8 = ___	9 + 8 = ___	9 + 3 = ___	9 + 6 = ___	9 + 7 = ___
9 + 2 = ___	9 + 1 = ___	9 + 8 = ___	9 + 9 = ___	9 + 3 = ___
9 + 4 = ___	9 + 6 = ___	9 + 7 = ___	9 + 2 = ___	9 + 5 = ___
9 + 9 = ___	9 + 4 = ___	9 + 9 = ___	9 + 7 = ___	9 + 1 = ___
Score: ____/25 _____ Min. _____ Sec.	Score: ____/25 _____ Min. _____ Sec.	Score: ____/25 _____ Min. _____ Sec.	Score: ____/25 _____ Min. _____ Sec.	Score: ____/25 _____ Min. _____ Sec.

Extra Practice Nine Plus Drills

Name: _____

Day 1	Day 2	Day 3	Day 4	Day 5
9 + __ = 12	__ + 2 = 11	__ + 6 = 15	9 + __ = 10	__ + 4 = 13
__ + 5 = 14	9 + __ = 13	9 + __ = 13	__ + 6 = 15	9 + __ = 9
9 + __ = 17	__ + 7 = 16	__ + 0 = 9	9 + __ = 18	__ + 8 = 17
__ + 2 = 11	9 + __ = 12	9 + __ = 14	__ + 2 = 11	9 + __ = 16
9 + __ = 13	__ + 5 = 14	__ + 2 = 11	9 + __ = 16	__ + 9 = 18
__ + 0 = 9	9 + __ = 10	9 + __ = 16	__ + 5 = 14	9 + __ = 11
9 + __ = 15	__ + 8 = 17	__ + 9 = 18	9 + __ = 12	__ + 6 = 15
__ + 9 = 18	9 + __ = 9	9 + __ = 12	__ + 8 = 17	9 + __ = 13
9 + __ = 16	__ + 9 = 18	__ + 8 = 17	9 + __ = 13	__ + 1 = 10
__ + 1 = 10	9 + __ = 15	9 + __ = 10	__ + 0 = 9	9 + __ = 14
9 + __ = 12	__ + 2 = 11	__ + 6 = 15	9 + __ = 10	__ + 0 = 9
__ + 5 = 14	9 + __ = 13	9 + __ = 13	__ + 6 = 15	9 + __ = 12
9 + __ = 17	__ + 7 = 16	__ + 0 = 9	9 + __ = 18	__ + 8 = 17
__ + 2 = 11	9 + __ = 12	9 + __ = 11	__ + 2 = 11	9 + __ = 14
9 + __ = 13	__ + 5 = 14	__ + 4 = 13	9 + __ = 16	__ + 9 = 18
__ + 0 = 9	9 + __ = 10	9 + __ = 9	__ + 5 = 14	9 + __ = 11
9 + __ = 15	__ + 8 = 17	__ + 5 = 14	9 + __ = 12	__ + 6 = 15
__ + 9 = 18	9 + __ = 9	9 + __ = 11	__ + 8 = 17	9 + __ = 13
9 + __ = 10	__ + 6 = 15	__ + 7 = 16	9 + __ = 13	__ + 1 = 10
__ + 3 = 12	9 + __ = 11	9 + __ = 18	__ + 0 = 9	9 + __ = 14
9 + __ = 14	__ + 4 = 13	__ + 3 = 12	9 + __ = 10	__ + 0 = 9
__ + 8 = 17	9 + __ = 16	9 + __ = 17	__ + 6 = 15	9 + __ = 12
9 + __ = 11	__ + 3 = 12	__ + 1 = 10	9 + __ = 18	__ + 8 = 17
__ + 4 = 13	9 + __ = 14	9 + __ = 15	__ + 2 = 11	9 + __ = 16
9 + __ = 18	__ + 1 = 10	__ + 4 = 13	9 + __ = 16	__ + 9 = 18
Score: _____ /25	Score: _____ /25	Score: _____ /25	Score: _____ /25	Score: _____ /25
_____ Min.	_____ Min.	_____ Min.	_____ Min.	_____ Min.
_____ Sec.	_____ Sec.	_____ Sec.	_____ Sec.	_____ Sec.

Nine Plus Review Test

Name: _____

9 +3	9 +5	9 +8	9 +2	9 +4	9 +0	9 +6	9 +9	9 +7	9 +1
9 +5	9 +9	9 +6	9 +3	9 +0	9 +2	9 +1	9 +8	9 +3	9 +4
9 +2	9 +2	9 +4	9 +5	9 +7	9 +4	9 +9	9 +3	9 +5	9 +8
9 +1	9 +8	9 +9	9 +0	9 +2	9 +6	9 +4	9 +7	9 +3	9 +5
9 +7	9 +3	9 +1	9 +4	9 +8	9 +5	9 +0	9 +1	9 +6	9 +2
9 +6	9 +4	9 +0	9 +5	9 +2	9 +7	9 +9	9 +3	9 +8	9 +1
9 +2	9 +5	9 +6	9 +7	9 +3	9 +4	9 +8	9 +0	9 +9	9 +4
9 +1	9 +4	9 +1	9 +0	9 +9	9 +2	9 +7	9 +5	9 +6	9 +3
9 +8	9 +0	9 +2	9 +6	9 +7	9 +1	9 +5	9 +3	9 +8	9 +4
9 +7	9 +8	9 +9	9 +5	9 +4	9 +2	9 +4	9 +6	9 +1	9 +0

Date: _____ Score: _____/100 Time: _____ Min. _____ Sec.

6+, 7+, 8+, 9+ Timed Drill Review

Name: _____

Row 1	Row 2	Row 3	Row 4
6 + 3 = _____	8 + 6 = _____	9 + 1 = _____	7 + 2 = _____
7 + 5 = _____	9 + 4 = _____	6 + 6 = _____	8 + 4 = _____
8 + 8 = _____	6 + 0 = _____	7 + 9 = _____	9 + 7 = _____
9 + 2 = _____	7 + 5 = _____	8 + 2 = _____	6 + 3 = _____
6 + 4 = _____	8 + 2 = _____	9 + 7 = _____	7 + 5 = _____
7 + 0 = _____	9 + 7 = _____	6 + 5 = _____	8 + 1 = _____
8 + 6 = _____	6 + 9 = _____	7 + 3 = _____	9 + 8 = _____
9 + 9 = _____	7 + 3 = _____	8 + 8 = _____	6 + 0 = _____
6 + 7 = _____	8 + 8 = _____	9 + 4 = _____	7 + 9 = _____
7 + 1 = _____	9 + 1 = _____	6 + 0 = _____	8 + 6 = _____
8 + 3 = _____	6 + 6 = _____	7 + 1 = _____	9 + 2 = _____
9 + 5 = _____	7 + 4 = _____	8 + 6 = _____	6 + 4 = _____
6 + 8 = _____	8 + 0 = _____	9 + 9 = _____	7 + 7 = _____
7 + 2 = _____	9 + 2 = _____	6 + 2 = _____	8 + 3 = _____
8 + 4 = _____	6 + 4 = _____	7 + 7 = _____	9 + 5 = _____
9 + 0 = _____	7 + 0 = _____	8 + 5 = _____	6 + 1 = _____
6 + 6 = _____	8 + 5 = _____	9 + 3 = _____	7 + 8 = _____
7 + 9 = _____	9 + 2 = _____	6 + 8 = _____	8 + 0 = _____
8 + 1 = _____	6 + 7 = _____	7 + 4 = _____	9 + 6 = _____
9 + 3 = _____	7 + 9 = _____	8 + 0 = _____	6 + 2 = _____
6 + 5 = _____	8 + 3 = _____	9 + 4 = _____	7 + 4 = _____
7 + 8 = _____	9 + 8 = _____	6 + 6 = _____	8 + 7 = _____
8 + 2 = _____	6 + 1 = _____	7 + 9 = _____	9 + 3 = _____
9 + 4 = _____	7 + 6 = _____	8 + 2 = _____	6 + 5 = _____
6 + 9 = _____	8 + 4 = _____	9 + 7 = _____	7 + 1 = _____

Date: _____ Score: _____/100 Time: _____ Min. _____ Sec.

OTM-1139 • SSK1-39 Timed Addition Facts

6+, 7+, 8+, 9+ Timed Drill Review

Name: _____

6 + 3	7 + 5	8 + 8	9 + 2	6 + 4	7 + 0	8 + 6	9 + 9	6 + 7	7 + 1
8 + 3	9 + 5	6 + 8	7 + 2	8 + 4	9 + 0	6 + 6	7 + 9	8 + 1	9 + 3
6 + 5	7 + 8	8 + 2	9 + 4	6 + 9	7 + 2	8 + 4	9 + 7	6 + 3	7 + 5
8 + 1	9 + 8	6 + 0	7 + 9	8 + 6	9 + 2	6 + 4	7 + 7	8 + 3	9 + 5
6 + 1	7 + 8	8 + 0	9 + 6	6 + 2	7 + 4	8 + 7	9 + 3	6 + 5	7 + 1
8 + 6	9 + 4	6 + 0	7 + 5	8 + 2	9 + 7	6 + 9	7 + 3	8 + 8	9 + 1
6 + 6	7 + 4	8 + 0	9 + 2	6 + 4	7 + 0	8 + 5	9 + 2	6 + 7	7 + 9
8 + 3	9 + 8	6 + 1	7 + 6	8 + 4	9 + 1	6 + 6	7 + 2	8 + 6	9 + 7
6 + 5	7 + 3	8 + 8	9 + 4	6 + 0	7 + 1	8 + 6	9 + 9	6 + 2	7 + 7
8 + 5	9 + 3	6 + 8	7 + 4	8 + 0	9 + 1	6 + 6	7 + 9	8 + 2	9 + 7

Date: _____ **Score:** _____ /100 **Time:** _____ Min. _____ Sec.

Review Drill of Addition Facts 0 to 18

Name: _____

A	4 +3	4 +4	3 +1	7 +1	5 +8	0 +2	2 +1	7 +4	1 +3	4 +5
B	3 +2	7 +3	2 +6	6 +6	8 +0	1 +2	8 +6	4 +4	8 +9	7 +7
C	1 +8	4 +1	7 +2	0 +6	7 +9	2 +8	8 +8	5 +9	2 +0	9 +3
D	3 +7	3 +4	5 +3	0 +5	9 +7	8 +1	5 +7	1 +7	8 +4	6 +1
E	4 +7	8 +2	0 +8	6 +3	4 +2	3 +5	9 +0	1 +1	7 +6	5 +4
F	0 +9	6 +4	4 +0	3 +6	1 +9	7 +0	6 +7	2 +4	9 +2	0 +7
G	9 +4	2 +2	5 +5	1 +5	6 +9	7 +5	2 +5	9 +6	0 +4	3 +0
H	2 +3	0 +3	9 +8	4 +8	6 +0	5 +6	3 +3	5 +1	0 +1	9 +5
I	0 +0	8 +5	4 +6	6 +2	1 +0	8 +7	5 +2	2 +7	6 +8	3 +8
J	4 +9	8 +3	2 +9	9 +1	3 +9	6 +5	1 +6	7 +8	9 +9	5 +0

Date: _____   Score: _____/100   Time: _____ Min. _____ Sec.

Review Drill of Addition Facts 0 to 18

Name: _____

A	5 +5	1 +5	9 +7	4 +7	2 +1	7 +3	0 +4	8 +4	6 +6	3 +7
B	0 +3	8 +9	5 +1	3 +0	9 +8	5 +4	2 +2	9 +6	7 +4	4 +8
C	7 +2	4 +5	2 +6	6 +7	1 +6	9 +2	3 +8	8 +5	0 +9	6 +1
D	5 +9	2 +8	7 +6	3 +9	8 +6	0 +8	6 +3	3 +1	0 +7	8 +3
E	4 +4	2 +0	6 +3	4 +6	1 +4	9 +5	7 +5	5 +3	8 +7	4 +3
F	1 +1	5 +8	8 +8	2 +6	3 +5	3 +7	1 +3	7 +4	9 +2	0 +0
G	8 +2	3 +6	7 +1	3 +4	1 +9	7 +8	6 +5	3 +5	0 +7	4 +2
H	5 +7	0 +5	6 +9	4 +1	3 +3	2 +2	6 +7	5 +2	6 +4	8 +1
I	6 +4	5 +5	6 +2	9 +1	2 +3	8 +0	6 +6	1 +1	0 +9	4 +9
J	4 +0	1 +6	4 +4	9 +9	2 +9	5 +6	3 +7	8 +1	6 +7	9 +5

Date: _____ Score: _____/100 Time: _____ Min. _____ Sec.

OTM-1139 • SSK1-39 Timed Addition Facts

Review Drill of Addition Facts 0 to 18

Name: _____

A	4 + 5 = ___	2 + 9 = ___	6 + 9 = ___	1 + 7 = ___	4 + 1 = ___
B	7 + 8 = ___	5 + 1 = ___	4 + 6 = ___	7 + 2 = ___	3 + 4 = ___
C	7 + 1 = ___	8 + 7 = ___	3 + 0 = ___	6 + 3 = ___	0 + 4 = ___
D	9 + 6 = ___	1 + 1 = ___	4 + 4 = ___	4 + 7 = ___	9 + 1 = ___
E	4 + 3 = ___	7 + 0 = ___	0 + 8 = ___	6 + 6 = ___	5 + 8 = ___
F	8 + 4 = ___	9 + 2 = ___	8 + 8 = ___	8 + 2 = ___	4 + 8 = ___
G	1 + 2 = ___	2 + 5 = ___	2 + 0 = ___	6 + 1 = ___	1 + 6 = ___
H	9 + 9 = ___	6 + 5 = ___	5 + 3 = ___	3 + 6 = ___	6 + 8 = ___
I	3 + 9 = ___	2 + 1 = ___	2 + 3 = ___	4 + 2 = ___	9 + 4 = ___
J	5 + 6 = ___	8 + 0 = ___	0 + 6 = ___	9 + 3 = ___	6 + 4 = ___
K	3 + 5 = ___	6 + 7 = ___	1 + 4 = ___	8 + 1 = ___	5 + 2 = ___
L	1 + 3 = ___	9 + 7 = ___	8 + 9 = ___	5 + 9 = ___	8 + 3 = ___
M	6 + 0 = ___	0 + 0 = ___	8 + 5 = ___	2 + 7 = ___	1 + 8 = ___
N	1 + 9 = ___	5 + 7 = ___	1 + 5 = ___	7 + 6 = ___	2 + 2 = ___
O	5 + 0 = ___	3 + 3 = ___	9 + 8 = ___	0 + 1 = ___	7 + 3 = ___
P	4 + 9 = ___	0 + 3 = ___	3 + 8 = ___	1 + 0 = ___	7 + 9 = ___
Q	7 + 5 = ___	2 + 4 = ___	7 + 7 = ___	3 + 1 = ___	0 + 2 = ___
R	0 + 9 = ___	5 + 5 = ___	4 + 0 = ___	2 + 6 = ___	8 + 6 = ___
S	6 + 2 = ___	3 + 2 = ___	9 + 5 = ___	9 + 0 = ___	3 + 7 = ___
T	0 + 5 = ___	7 + 4 = ___	2 + 8 = ___	0 + 7 = ___	5 + 4 = ___

Date: _____   Score: _____/100   Time: _____ Min. _____ Sec.

Review Drill of Addition Facts 0 to 18

Name: _____

A	3 + 0 = ___	6 + 2 = ___	1 + 4 = ___	2 + 0 = ___	5 + 4 = ___
B	9 + 6 = ___	5 + 1 = ___	8 + 2 = ___	9 + 2 = ___	8 + 0 = ___
C	2 + 0 = ___	2 + 9 = ___	6 + 7 = ___	7 + 2 = ___	0 + 4 = ___
D	7 + 4 = ___	1 + 0 = ___	9 + 9 = ___	3 + 4 = ___	8 + 5 = ___
E	4 + 2 = ___	3 + 8 = ___	0 + 7 = ___	9 + 4 = ___	1 + 5 = ___
F	1 + 3 = ___	8 + 7 = ___	3 + 3 = ___	0 + 1 = ___	7 + 6 = ___
G	4 + 5 = ___	2 + 1 = ___	2 + 8 = ___	7 + 0 = ___	5 + 3 = ___
H	2 + 2 = ___	1 + 6 = ___	4 + 1 = ___	7 + 9 = ___	6 + 1 = ___
I	1 + 7 = ___	5 + 7 = ___	6 + 0 = ___	4 + 6 = ___	3 + 5 = ___
J	0 + 6 = ___	8 + 8 = ___	1 + 2 = ___	6 + 8 = ___	5 + 2 = ___
K	5 + 9 = ___	5 + 6 = ___	2 + 3 = ___	0 + 8 = ___	3 + 7 = ___
L	7 + 8 = ___	9 + 5 = ___	5 + 0 = ___	4 + 4 = ___	9 + 7 = ___
M	9 + 1 = ___	0 + 0 = ___	7 + 5 = ___	2 + 6 = ___	1 + 1 = ___
N	4 + 0 = ___	6 + 5 = ___	6 + 4 = ___	3 + 6 = ___	4 + 8 = ___
O	8 + 3 = ___	3 + 9 = ___	9 + 3 = ___	4 + 7 = ___	7 + 1 = ___
P	2 + 5 = ___	7 + 7 = ___	0 + 5 = ___	8 + 6 = ___	2 + 7 = ___
Q	1 + 9 = ___	9 + 8 = ___	8 + 4 = ___	0 + 3 = ___	8 + 1 = ___
R	3 + 2 = ___	4 + 3 = ___	4 + 9 = ___	6 + 9 = ___	5 + 5 = ___
S	7 + 3 = ___	6 + 6 = ___	9 + 0 = ___	3 + 1 = ___	8 + 9 = ___
T	5 + 8 = ___	1 + 8 = ___	2 + 4 = ___	6 + 3 = ___	0 + 9 = ___

Date: _____   Score: _____/100   Time: _____ Min. _____ Sec.

OTM-1139 • SSK1-39 Timed Addition Facts

Score Record Sheet for _____ **Drills**
Name: _____

1. Date: _____
   Score: ___/25 Time: ___Min. ___Sec.

2. Date: _____
   Score: ___/25 Time: ___Min. ___Sec.

3. Date: _____
   Score: ___/25 Time: ___Min. ___Sec.

4. Date: _____
   Score: ___/25 Time: ___Min. ___Sec.

5. Date: _____
   Score: ___/25 Time: ___Min. ___Sec.

6. Date: _____
   Score: ___/25 Time: ___Min. ___Sec.

7. Date: _____
   Score: ___/25 Time: ___Min. ___Sec.

8. Date: _____
   Score: ___/25 Time: ___Min. ___Sec.

9. Date: _____
   Score: ___/25 Time: ___Min. ___Sec.

10. Date: _____
    Score: ___/25 Time: ___Min. ___Sec.

11. Date: _____
    Score: ___/25 Time: ___Min. ___Sec.

12. Date: _____
    Score: ___/25 Time: ___Min. ___Sec.

My speed and accuracy is

_____ .

Score Record Sheet for _____ **Drills**
Name: _____

1. Date: _____
   Score: ___/100 Time: ___Min. ___Sec.

2. Date: _____
   Score: ___/100 Time: ___Min. ___Sec.

3. Date: _____
   Score: ___/100 Time: ___Min. ___Sec.

4. Date: _____
   Score: ___/100 Time: ___Min. ___Sec.

5. Date: _____
   Score: ___/100 Time: ___Min. ___Sec.

6. Date: _____
   Score: ___/100 Time: ___Min. ___Sec.

7. Date: _____
   Score: ___/100 Time: ___Min. ___Sec.

8. Date: _____
   Score: ___/100 Time: ___Min. ___Sec.

9. Date: _____
   Score: ___/100 Time: ___Min. ___Sec.

10. Date: _____
    Score: ___/100 Time: ___Min. ___Sec.

11. Date: _____
    Score: ___/100 Time: ___Min. ___Sec.

12. Date: _____
    Score: ___/100 Time: ___Min. ___Sec.

My speed and accuracy is

_____ .